Evidence Ensembles

Edited by
Christoph Rosol, Giulia Rispoli, Katrin Klingan,
and Niklas Hoffmann-Walbeck

T0052652

EVIDENCE ENSEMBLES   Edited by Christoph Rosol, Giulia Rispoli, Katrin Klingan, and Niklas Hoffmann-Walbeck, with contributions by Kat Austen, Carbon Aesthetics Group (Desiree Foerster, Myriel Milićević, Karolina Sobecka, Alexandra Regan Toland, Clemens Winkler), Nigel Clark, Kristine L. DeLong, Anna Echterhölter, Victor Galaz, Susan Schuppli, Olúfẹ́mi O. Táíwò, Liz Thomas, Simon Turner, Abraham Gottlob Werner, and Jens Zinke

Markers that Matter

When geologists unearth the more recent layers in the Earth's archives, they encounter distinct material traces of human impacts on a global scale. Throughout the still relatively shallow sediment of human history, anthropogenic residues appear in many places and at various depths; they are scattered, indicative of various levels of intensity, and diachronic—that is, they imply changes over long periods of time. Such archaeological or historical evidence, however, is usually confined to local settings and settlements or constitutes a low-amplitude signal within the background noise of natural processes. But the picture changes radically towards the upper ends of the core samples that show depositional residues from the twentieth century. Here, the anthropogenic signals become much more pronounced, diverse, and globally similar, providing a material account of a highly dynamic transformation of not only a local site, but of the entire planet.

The art of studying earthly material evidence and its relationship to the layering of geological time is called chronostratigraphy. The chronostratigraphic reality of the Anthropocene just described has been examined by the Anthropocene Working Group (AWG), an international body of Earth scientists formed by the International Commission on Stratigraphy in 2009. Together with a number of geological teams from around the world, AWG researchers have been looking for the measurable impact of global human activities in lake sediments, corals, speleothems, a peat bog, an ice shelf, and various other geological archives. Interestingly, and for the first time in their disciplinary history, these chronostratigraphers were now faced with the challenging task of pinpointing a geological epoch whose beginning lies within the lifetime of their own parents' generation. They had to conduct a "geology of the present," a structural analysis of that which is still in the process of unfolding.

In effect, our planet is experiencing only the first stage of the Anthropocene: a highly disruptive transitional period of "global weirding" in which established ecological, climatic, geochemical, and biological patterns are changing radically, and in some cases are threatened with collapse. The material evidence for this planetary transition is abundant and mounting. The registration of this evidence is of immense importance for understanding that we are indeed leaving behind an epoch—that of the Holocene—and that we are closing the window to a certain way of relating to the planet based on an ever-accelerating exploitation of its resources and devastation of its natural tissue. For future habitability, other ways of relating need to be explored and developed, new cosmologies that allow for an understanding of interdependency and interspecies care, so that other material evidence—born from the aptitude of humans to learn and correct themselves—may start to accumulate.

Evidence Ensembles presents an experimental collage of perspectives and narrations around the material witnesses of planetary change. The contributions and discursive constellations in this volume set up a conversation among geologists, historians, philosophers, and artists. Different perspectives inform each other through the registers of particular matters, epistemic questions, and shared concerns. How does humankind's global fingerprint appear in earthly strata? How are human materials classified, and how do human structures compare to biological ones? What confines temporal data and what politicizes earthly matters?

Victor Galaz, a political scientist, and Simon Turner, an environmental scientist and secretary of the AWG, open this collection with their reflections on the material and immaterial traces that highlight the transformational effects of humans on the planet. How is the volatility of information in the geological record best captured? The search for circumstantial evidence depends on many factors: one does not always have an objective document to decipher and what survives is not the totality of that which existed in the past.

Still, there is nothing immaterial about this process. The essay by historian of science Anna Echterhölter perambulates the fascinating domain of anthropogenic minerals and technofossils, novel substances, which now sweep away the traditional classificatory distinctions between pure matter and the plant and animal kingdoms and as such require conceptual innovation with regard to these emerging entities. She shows that mineral classification, understood in the Western canon as something detached from human agency, is, in fact, deeply entangled with it. Scientific taxonomies, therefore, reveal certain norms and value systems and the policies that have been established to consolidate these.

Along these lines, Victor Galaz and Simon Turner return to the question of who actually defines the Anthropocene and whether the stratigraphic framework used to determine it might need to be revised in light of how this particular concept goes far beyond regular scientific interest in periodization. This aspect remains thorny. The AWG's work, by definition, has to be conservative and stick to the protocols of chronostratigraphy. On the other hand, the AWG's commendable engagement in a continued transdisciplinary dialogue—as the contributions in this volume demonstrate—has opened up new spaces for addressing the markers and matters of the Anthropocene.

Conceptual, if not ontological divisions are once again dissolved in the conversation between artist Kat Austen, human geographer Nigel Clark, and paleoscientists Kristine DeLong and Jens Zinke as they compare the similarities of natural and cultural structures. Their protagonists are corals, lush biogenic colonies that, like human habitats, erect themselves into three-dimensional space, house other co-dwellers, and migrate to other places as climatic conditions change.

Like corals, ice sheets hold astonishing information about the history of Earth's mobile matters such as chemical compounds or dust. Two experts on ice, paleoclimatologist Liz Thomas and artist-researcher Susan Schuppli characterize this curious material as a custodian of a multiplicity of temporalities that coexist and overlap. In a second outtake of their conversation, they continue discussing the quality of Antarctic ice to store the extent of past sea ice and the direction and strength of past winds. Yet, long before Antarctica became a peaceful storehouse of geoscientific information in the mid-twentieth century, the white continent also recorded another type of information, as Liz Thomas reports: the historical extent of the whaling industry in the nineteenth century.

Whales are indeed the central characters of a further essay in this collection in which the artistic research group Carbon Aesthetics construes these creatures as emblematic of the shifting registers between humans and nonhumans. Whales make material interrelations traceable over time and watery spaces—as sensor and signifier for the cruelties of the whaling industry, the boons of illumination of human habitats before fossil fuels, and bodies of "no tech" carbon storage for the carbon capture economy of today. From the eighteenth century until now, whales tell a story of visibility, and hence the perception of the world through modern aesthetics of carbonaceous industrialization.

In a second part of their conversation, Kat Austen, Nigel Clark, Kristine DeLong, and Jens Zinke ponder the depth of time implicit in coral evolution and what these creatures might be telling us about our own future. Dieback now threatens coral cities just as rising sea-levels threaten human-built cities on Holocene coastlines or islands. Will their sunken remains once again become reefs for surviving corals?

The quintessential element of human civilization—and its possible sinking—is the access to and control of boundless energy, and so Nigel Clark's essay sheds light on a crucial phenomenon of modern humans' existence: anthropogenic fire. Clark looks at the ways by which fire has been captured, controlled, and inverted: from open landscape burning to the chambered combustion in heat engines to the near-instantaneous combustion during the detonation of weaponized explosives. In that pyric transition, the burning of fossil reserves ignites, indeed, a significant stage of fire control, one that manifests the blazing mediation between geological strata and planetary processes.

As these perspectives show, histories of anthropogenic markers extend understanding of the Anthropocene both within and beyond stratigraphic evidence, highlighting cultural formations as complementary to geological ones. In a final excerpt of their conversation, Victor Galaz and Simon Turner deal with the question of how far the geology of the Anthropocene can unravel societal and political processes. What then is the implicit political narrative of a formalized Anthropocene?

In the concluding essay, Olúfẹ́mi O. Táíwò proposes the slow formation and violent destruction of cave structures—the stalagmites and stalactites that the AWG has also included in their analysis—as a metaphor for the slash-and-burn pathway of human history. Just as the slow accumulation of mineral deposits make up the morphology of a cave, the violence of European colonialism, which began its spread across the globe in the fifteenth century, slowly solidified into the institutions, traditions, and norms that now make up the "global racial empire." Táíwò's contribution is a clarion call to finally abolish these cavernous structures of exploitation and suppression.

The images scattered throughout the pages of this volume represent a historical attempt at a chromatic ensemble of earthly evidence. The color-plates of Meissen porcelain are part of a collection of 249 plates that the German mineralogist Abraham Gottlob Werner fabricated at the beginning of the nineteenth century. The nomenclature developed by Werner was part of his theory of the characteristics of fossils, which was intended to facilitate the unambiguous description and identification of minerals by means of sensory perception. In her contribution to this volume, Anna Echterhölter contextualizes

Werner's classification project in more detail. The porcelain color-plates displayed here reflect not only the chromatic design of the DNA books' series but also remind us of the historically changing and at the same time eternally renewing connection between knowledge and method that underlies every form of sampling and analyzing Earth's material evidence.

The essays and conversations presented in this volume comprise a selection of contributions from the online publication *Anthropogenic Markers: Stratigraphy and Context,* a project by the Max Planck Institute for the History of Science (MPIWG), and edited excerpts from conversations held during the *Unearthing the Present* event that took place at Haus der Kulturen der Welt (HKW) over the course of May 19–22, 2022. Both resources are the result of the joint, two-year working project of the MPIWG, HKW, and AWG to contextualize and expand upon the final investigations of the AWG that set out to formalize the Anthropocene epoch. The material generated by this project and further information about it can be found on anthropocene-curriculum.org as well as in *The New Alphabet, Volume 17: Geology of the Present*.

Christoph Rosol, Giulia Rispoli, Katrin Klingan, and Niklas Hoffmann-Walbeck

Rossen. Purp.

69.

Material and Immaterial Traces

Victor Galaz: Many of the most important things in societies do not leave clear material traces, but they still shape our planet and result in things like norms and social inequality. What counts as material evidence, and how do you deal with events or social changes that do not leave a direct material trace in the ground?

Simon Turner: You must always think about what is not in the geological record. An awful lot of material you find in the geological record only represents a very small amount of what has actually happened. Take the Chicago Stock Exchange for instance, and all the information on finance; the work practices and codes may suddenly change, with the digitalization of commerce for example, and all previous requirements become redundant, disappearing from the record. This kind of information disappears very quickly in the geological/archaeological record, because how do you record information geologically unless you have artifacts or physical evidence?

Once you go past a certain point—past history, past archaeology—you are left with just physical materials, and they are quite difficult to interpret. Say you look at a certain kind of microscopic particle involved in power production, which is clear evidence of an industrial process and how that process changes over time. As their abundance in a sediment sequence increases, you can observe that there was a change in production, so you can start inferring that power production was increasing. Why would you have an increase in power production? Probably because people wanted and used more power. Then you can say, well, it was probably a political or social structural change in how society works, which is then reflected in the profile of a particle pollutant.

But I think that's probably as far as you can go—inferring. In archaeology, rarely can you definitively say, "this find represents a change from one taxation system to another"; for that, something usually has to be written down, you have to have documentary evidence. But as soon as you lose that documentary evidence or that social evidence or the written word, then it really becomes a matter of inference. At some point, information is lost. And that is kind of an interesting thing for us, moving towards an information society. I mean, I worry about my USB sticks.

VG: Think about a future AWG [Anthropocene Working Group] in, let's say, 300 years' time, and imagine that our societies collapse. So, looking back in 300 years, you would not be able to see the vast digital connections of right now, you would probably just see the resulting consumption of resources, but not other aspects and more important social aspects.

ST: Well, you would still be able to infer that something was definitely going on. And you would have to infer based on what we—us in 300 years—know about how power was created. So, alongside other information gathered, we would still be able to infer things from geological evidence. Unless it is an information-free society in 300 years, which is probably not going to happen. We will still need information; that is probably a guarantee.

VG: But is that not a bit of a challenge for you today? The fact that people are expecting the AWG to not only say something about geological changes, but also about the drivers behind these changes.

ST: Yes, and I also think that's why the Anthropocene concept has been picked up so widely across disciplines. Perhaps people are asking a little bit too much of geology: to say everything about anthropogenic drivers. We can identify stratigraphically the increase in CO_2 in the atmosphere or the change in nitrogen composition, and we can say that there is pretty good evidence that a change in the presence of an artificial fertilizer is related to its invention and application. So, there are big societal drivers we can identify, but we cannot say it was definitely President Roosevelt who did this or whether it was that one political decision in China that clearly created this environmental signal that has been recorded stratigraphically.

We are looking at the planetary scale and things that are recorded on that scale. So, tying it back to specific social drivers is difficult—and being more definite is a challenge.

Take plutonium for instance. We can say that a change in isotope stratigraphy is the result of a specific Pacific atomic bomb test. And then you can link that back to a series of detonations, which would be, say, definitely related to the United States rather than France. With this you could say, "well, there's a clearly political thing. There is a clear driver." But for a lot of these markers, it's too global to tie into something specific. I guess that's where we hand over to the political and social scientists to see what they think about which changes were large enough to have driven the geological conditions. We need both perspectives.

The above conversation is an edited excerpt from "Exchange on Collaboration and Complexity," a discussion held on May 21, 2022 at HKW in Berlin during the event *Unearthing the Present*.

Human–Mineral Classification: Taxonomy, Totemism, and the Technofossils of the Anthropocene

Technofossils and a Twofold Disarray

Since the mid-twentieth century, matter has been becoming less and less "natural." Geological forces are no longer the sole drivers of the formation of solid substances. A term suggested by geologists for these new formats of "earth" is "technofossils." They have begun to cover the globe and constitute its most recent strata. Among them are chemical artifacts like boron nitride and tungsten carbide as well as "mineraloids" such as artificial glasses and plastics.[1] Likewise, defunct products factor into this category, from kitchen appliances to concrete rubble. "Technofossils" may be mistaken as an attention-seeking new label for the by-products of the industrial age or be seen as a first-rate taxonomic provocation. The suggestion laid out in the following paragraphs is to instead think of technofossils as a first-rate *conceptual innovation*.

Due to their reliable presence, technofossils are part and parcel of Western science's conception of material and earthly matters. But while natural minerals could be considered the best examples of "matter" or "physics" around—they are visible, tangible, stable, and hence obviously real—these new minerals and compounds only pretend to be natural matter, and geological forces alone cannot explain their existence.[2] For example, minerals are, by definition, formed by

---

1   Jan Zalasiewicz, Mark William, Colin N. Waters, et al., "The Technofossil Record of Humans," *Anthropocene Review*, vol. 1, no. 1 (2014), pp. 34–43, here p. 36.

2   Jan Zalasiewicz, Colin N. Waters, Erie C. Ellis, et al., "The Anthropocene: Comparing Its Meaning in Geology (Chronostratigraphy) with Conceptual Approaches Arising in Other Disciplines," *Earth's Future*, vol. 9, no. 3 (2021), https://agupubs.onlinelibrary.wiley.com/doi/full/10.1029/2020EF001896; Emily Elhacham, Liad Ben-Uri, Jonathan Grozovski, et al., "Global Human-Made Mass Now Exceeds All Living Biomass," *Nature*, vol. 588 (December 2020), pp. 442–44; Robert M. Hazen, Edward S. Grew, Marcus J. Origlieri, et al., "Mineral Evolution," *American Mineralogist*, vol. 93, no. 11/12 (2008), pp. 1693–720. All online references in this essay were last accessed in October 2022.

geological forces—and yet around 5,100 recognized mineral specimens occur in artificial settings like mines. Around 200 accepted and
ratified mineral specimens like delrioite, schuetteite, and widgiemool-
thalite do not emerge from "natural" processes at all. Thus, quite a
few mineral entities have to be considered as at least partly synthetic.[3]
Within the framework of traditional geology, it remains odd that minerals never known to have been produced by geological forces occupy
the same classificatory space as rocks and metals.

From the perspective of chronostratigraphy, technofossils func-
tion in much the same way as fossilized organisms, which occur only
in select geological strata. They are occurrences in nature, have a factual, material existence, and offer important signals. Just as the beginning of the Devonian period is indicated by fossils of a new species
(*Monograptus uniformis*), technofossils may be what's chosen to mark
the onset of the Anthropocene. Whatever the case, this class of materials can be considered the expression of "the geology of mankind"[4]
on the level of mineral classification.

But it upsets this order at the same time. Technofossils confound
the three kingdoms of nature (animal, vegetable, mineral) and constitute an awkward artificial-mineral-human realm. And it is this seem-
ing disarray that may prove to be a truly radical innovation. Crucially,
this nongeological matter brings to the fore the extent to which our
perception of the physical world hinges on conceptions like organic
versus inorganic matter and the division between nature and artifice.
One cornerstone of modern mineralogy is the concept of simple substance, which describes a chemical element in its purest form. What
ultimately makes technofossils intriguing is the double provocation
they embody for the conceptual architecture of mineralogy. On the one
hand, "technofossils" seems to be a necessary term—rather factual
and descriptive of what is happening on the ground, in the physical

---

3   Robert M. Hazen, Dominic Papineau, Wouter Bleeker, et al., "On the
    Mineralogy of the 'Anthropocene Epoch,'" *American Mineralogist*,
    vol. 102, no. 2 (2017), pp. 595–611; Jeffrey de Fourestier, "The Naming
    of Mineral Species Approved by the Commission on New Minerals and
    Mineral Names of the International Mineralogical Association: A Brief
    History," *The Canadian Mineralogist*, vol. 40, no. 6 (2002), 1721–35.
4   Paul J. Crutzen, "Geology of Mankind," *Nature*, vol. 415, no. 23 (2002),
    https://doi.org/10.1038/415023a

world, on a global and massive scale.[5] Yet, the idea of the technofossil is at odds with the central conceptualization of the geological force of mineralogy, since technofossils are a result of human, not solely geological, production. This is the first disarray they cause. Technofossils also do not sit well with the descriptive terms customary to the geosciences. Reproach and even accusation resound in this new term, as if these modern fossils constituted a "wrong" thing, a step too far: technofossils are conceptualized as illegitimate and misplaced matter. Ultimately, it is impossible to conceive of technofossils without at least a faint echo of ecological scandal. This inherent normative element is the second dimension of technofossil disarray. Thinking with this concept entails delineating "what ought to be." Socioecological implications clearly constitute a breach in the tradition of mineralogical classification, which so far has kept people and the sphere of their actions out of the picture. Now the geological record is suggesting otherwise. Truly, though, technofossils amalgamate humankind and rocks in an awkward manner. This new group of mineral pretenders describes more than matter: it includes new ways of becoming nature, along with a certain sense of alarm.

To shed light on the twofold provocation of these mineral types (geological versus human-made, descriptive versus normative units), we'll now revisit two historical instances of mineral classification from the vantage point of the history of concepts.[6] The first case study looks at examples from Germany regarding the development of the simple substance concept over the course of the eighteenth century. It will establish the cornerstones of mineralogical classification and draw on research emphasizing the logics and practices of precise measurements of value, which form the immediate context of the "simplicity" or "purity" of a substance. The second case study returns to an episode from the history of ethnography. It touches upon an alternative classification of stones according to the totemic imaginations described by ethnographers in relation to several Pacific Islander societies. In both cases, we'll pay attention to the local and universal

5 Of course, the occurrence of technofossils is happening in unison with the expansion of the technosphere and its productions. See Elhacham "Global Human-Made Mass Now Exceeds All Living Biomass."

6 Henning Schmidgen, "The Life of Concepts: Georges Canguilhem and the History of Science," *History and Philosophy of the Life Sciences*, vol. 36, no. 2 (2014), pp. 232–53.

meanings of the scales involved, as they order minerals and situate their specific meanings within particular times and places on the globe; Anthropocene perspectives, of course, should always challenge us to consider such issues in terms of larger scale and specific locality.[7] Both revisions show how classificatory systems of matter depend greatly on particular socioeconomic practices. The latter explain specific modes of mineral classification in all cases more precisely than the introduction of the "human species" in general as characteristic of the new era of the Anthropocene.[8]

From this vantage point, it becomes less surprising that human agency and social concerns are invested in notions of what a mineral specimen is. Technofossils are not remarkable because, with a closer look, human–mineral relations can be deciphered. Rather, they are a conceptual innovation from within the context of the Western scientific tradition because they explicitly mingle human agency into geological classification and because they do so with loud normative claims. The scandal of transgressing the former confines constructed around human action resounds with their very existence. The normative overtones of technofossils portend the extinction of species occurring around us, and thus the classificatory term gestures implicitly toward a new and cautious version of economizing resources. The emphasis on "technofossils as conceptual innovation," spelled out in the following passages, complies with social historical concerns surrounding quantification that call for complex readings of apparently neutral concepts like metrics, data architectures, classificatory systems, concepts, standards, scales, and (mineralogical) units.

Simple Substance and Mineral Value

Brass and iron, bricks and sandstone—the sense of familiarity of these pairings derives from a history of smelting and construction practices.

7   Gabrielle Hecht, "Interscalar Vehicles for an African Anthropocene: On Waste, Temporality, and Violence," *Cultural Anthropology*, vol. 33, no. 1 (2018), pp. 109–41; Julia Adeney Thomas, "History and Biology in the Anthropocene: Problems of Scale, Problems of Value," *The American Historical Review*, vol. 119, no. 5 (2014), pp. 1587–607.
8   Andreas Malm and Alf Hornborg, "The Geology of Mankind? A Critique of the Anthropocene Narrative," *Anthropocene Review*, vol. 1, no. 1 (2014), pp. 62–69.

For a taxonomic point of view, these forms of matter are worlds apart. While iron and sandstone occur in nature, brass and bricks are human-made. Thus, we can see that artificial compounds match and rival those that are the product of geological forces, although some are more readily perceived as "natural" than others.

Even minerals, though, which are formed by the Earth, are rarely pure and simple enough in nature to live up to the mineralogical classes. Clearly, an overwhelming preponderance of "mixed" matter exists in nature. Even substances like gold or copper, which can be found in fairly pure states, are typically refined, further purified, and turned into more homogenous versions of themselves by humans. This raises the question as to why mineralogy began working with these idealized pure states in the first place, and where and in which contexts this core conception of mineralogical classification emerged.

Mineralogy, as a body of formalized scientific knowledge, consolidated in the West over the course of the eighteenth century. While simple kinds of mineral taxa surfaced in Swedish and German publications from the 1730s onward,[9] the impulse to purify can be traced way back to technological literature on assaying and hallmarking beginning in the 1550s.[10] As such, there is good reason to believe that the analytical definition of simple substance, which is key to chemical procedures, did not emerge from progress in chemistry or laboratory precision weighting alone, as some have suggested. Instead, this logic of "pure matter" can be strongly identified with the measurement of highly valuable ores—and so its history likely extends into mining, assaying, minting, and the administrative surveying of money issues, especially the policing of weighting systems related to coin production. This highly specialized area of precision was focused on the purity of

9   Of course, alternative traditions have also been described. See, for example, Silvia Fernanda de Mendonça Figueirôa, "The 'Table of Mineral Classification' by Oscar Nerval de Gouvêa: Mineralogy and Medicine in Brazil," *História, Ciências, Saúde-Manguinhos*, vol. 28, no. 2 (2021), pp. 491–508; Mikhail Yur Povarennykh, "The Crystal-Chemical Paradigm of the Modern Mineralogy (The Beginning of the XX Century-The Beginning of the XXI Century). What Is Next? Ontogenical Paradigm," *Urals Geological Journal*, vol. 111, no. 3 (2016), pp. 18-32.

10 Theodore M. Porter, "The Promotion of Mining and the Advancement of Science: The Chemical Revolution of Mineralogy," *Annals of Science*, vol. 38, no. 5 (1981), pp. 543-70.

gold, silver, and copper. Technical knowledge and ownership thus developed in lockstep. Where high value is involved, there is very good reason to quantify with accuracy and to monitor and consistently measure the purity of a substance.[11] And, indeed, we find that mineral classification using the notion of simple substance first occurs in books describing the art of assaying monetary metals.[12]

According to historian of science Theodore Porter, mineralogists and chemists such as Torbern Bergman and Antoine Lavoisier simply transferred this perspective from the mines and the assayer's workshop to the laboratories of chemistry. From here, purity, inseparability, and simplicity were imported into chemical analysis, as a practical logic, and thus transposed into the laboratory sphere. All this helped to practically and substantially arrive at the taxonomy of elements on which the mineral taxonomy came to be built.

Naming practices were one facet of this ongoing attempt to find states of matter that could not be separated any further, even by experimental means. Chemical analysis, and later crystallography, which classifies crystals according to their angles and internal structures, became more and more important in making sense of the sweeping variety of minerals that naturally occur in the Earth.

These developments, though, did have their adversaries. It was by no means an easy task to universalize the classificatory order of simple substance in practice. The very successful mineralogical school associated with Saxon mining is a case in point. At the end of

---

11 Ibid.
12 See Johann Andreas Cramer, *Elements of the Art of Assaying Metals*. London: T. Woodward, 1741 (09-18-2008), https://archive.org/details/ elementsofartofa00cram; Robert Siegfried and Betty Jo Dobbs, "Composition: A Neglected Aspect of the Chemical Revolution," *Annals of Science*, vol. 24, no. 4 (1968), pp. 275–93; Matthew D. Eddy, *The Language of Mineralogy: John Walker, Chemistry and the Edinburgh Medical School, 1750-1800*. Abingdon: Routledge, 2008; Martin Guntau, "Zu einigen Wurzeln der Mineralogie in der Geschichte," in Horst Kant and Annette Vogt (eds), *Aus Wissenschaftsgeschichte und -theorie Hubert Laitko zum 70. Geburtstag überreicht von Freunden, Kollegen und Schülern*. Berlin: Engel, 2005, pp. 111–31; Robert M. Hazen, "Mineralogy: A Historical Review," *Journal of Geological Education*, vol. 32, no. 5 (1984), pp. 288–98; Martin Rudwick, "Minerals, Strata and Fossils," in Nicholas Jardine, James A. Secord, and Emma C. Spary (eds), *Cultures of Natural History*. Cambridge: Cambridge University Press, 1996, pp. 266–86.

the eighteenth century, German aristocrats from Novalis to Alexander von Humboldt flocked to the remote mining town of Freiberg to hear the geologist Abraham Gottlob Werner explain the emergence of mountains (*Geognosie*) and to determine what constitutes minerals and rocks (*Oryktognosie*).[13] Werner's brand of mineralogy can be considered a stronghold against abstract and purified ways of classifying minerals, as he rejected criteria such as chemical composition and physical properties such as the magnetic or electrical behaviors of a substance. As the title of his influential classificatory textbook reveals,[14] Werner opted to focus on the "outward characteristics" of minerals—a decision usually explained by his aversion to identifying an arbitrary order in nature. He insisted instead on the "natural suite of bodies," which was tantamount to considering the mineral taxonomy to mirror natural order itself.[15] The quest to find a "natural" system of classification was a pursuit throughout the various branches of natural history.

Earlier explanations for Werner's refusal to switch to chemical analysis and the "inward characteristics" of modern science hinged on a philosophy of the senses, which, in his opinion, were the only organs necessary to arrive at the true order of mineralogy in its entirety and perfection.[16] All necessary information could and had to be obtained through outward criteria of color, weight, smell, taste, solidity, smoothness, coldness, and the like.[17] Furthermore, recent research

---

13 Samuel Gottlob Frisch, *Lebensbeschreibung Abraham Gottlob Werners. Nebst zwei Abhandlungen über Werners Verdienste um Oryktognosie und Geognosie von S. C. Weiß*. Leipzig: F. A. Brockhaus, 1825, p. 81 (n.d.), https://doi.org/10.3931/e-rara-67224

14 Werner's textbook is titled *Von den äußeren Kennzeichen der Fossilien*, or in English: *A Treatise on the Outward Characteristics of Minerals*.

15 Abraham Gottlob Werner, *Von den äusserlichen Kennzeichen der Fossilien*. Leipzig: Crusius, 1773, p. 21 (09-23-2008), https://deutsche-digitale-bibliothek.de/item/VMX6TUFAOJPO42AAEGORLFMOG7RAWZKW. The German expression reads "*natürliche Folge der Körper*."

16 "Denn ein Mineralsystem hat keinen andern Zweck, als die natürliche Folge oder Reihe der verschiedenen Fossilien zu bestimmen und je genauer dieses darinnen geschieht, je vollkommener wird das Mineralsystem sein" (p. 10); "bloß durch unsere Sinne" (p. 32), in Werner, *Von den äußeren Kennzeichen der Fossilien*.

17 Abraham Gottlob Werner, quoted from Martin Guntau, *Abraham Gottlob Werner*. Leipzig: Teubner, 1984, p. 95.

into the practices of mining have opened another and less phenomenological angle on the history of mineralogy, namely the fact that the
various professions that utilized this technology cultivated an unwrit-
ten, unformalized form of classification of matter according, precisely, to practical needs. Werner himself opted for a fivefold order of
"rock" only a few years before publishing his system of mineralogy,[18]
and this order corresponded with the five technologies then available
to process rock and exploit mining sites.[19] This "local" knowledge
with its practicability may have been yet a more mundane reason to
stick to outward criteria.

In fact, for Werner and his contemporaries, a switch to a system
based on simple substance and the internal criteria of minerals—requiring laboratory chemical analysis—would have meant making
mineralogy less accessible to general users, including amateur mineralogists and dilettantes, and narrowing its application in the field.[20]
These wider mineral audiences could practice their pastime much
more reliably by adhering to the minute color codes and sensory descriptions advocated by the field guides of the day, such as the one
published by Johann Georg Lenz, professor of mineralogy, in 1798.[21]
Apprehending and describing the exact graduation of sharpness or a
specific shade of a green that is milky and yellow at the same time was
challenging and required training.[22] While more accessible, determining outward characteristics nevertheless required an eye well-
trained in classificatory observation.

18 Abraham Gottlob Werner, *Von den verschiedenen Graden der Festigkeit des
Gesteins als dem Hauptgrunde der Hauptverschiedenheiten der
Häuerarbeiten.* Freiberg: Craz, 1788 (02-17-2016), https://archive.org/
details/bub_gb_CK6T7kgrVq8C
19 Sebastian Felten, "Wie fest ist das Gestein? Extraktion von
Arbeiterwissen im Bergbau des 18. Jahrhunderts," *WerkstattGeschichte*,
vol. 3 (2020), pp. 15–36; Ursula Klein, *Nützliches Wissen: Die Erfindung
der Technikwissenschaften.* Göttingen: Wallstein, 2016.
20 Porter, "The Promotion of Mining and the Advancement of Science,"
p. 548; Abraham Gottlob Werner, quoted from Frisch, *Lebensbeschreibung
Abraham Gottlob Werners*, p. 95.
21 Johann Georg Lenz, *Mineralogisches Taschenbuch für Anfänger und
Liebhaber*, 2 vols. Erfurt: Hennings, 1798/1799 (09-21-2009), https://
digitale-sammlungen.de/de/view/bsb10284059
22 Ibid.

Thus, the classificatory criteria chosen by Werner—one of the most influential experts of his time—proved to be motivated by a context of professional mining combined with a consideration of wealthy citizens and collectors interested in minerals. He seems to have had very good reasons to resist abstract classificatory criteria like those suggested by crystallography and chemistry, instead sticking to a schema based on workflows, the senses, and the outward characteristics of the mineral specimen. However, while practical mining concerns and the interests of collectors stood up to internal characteristics for a little while, it eventually became too difficult to disentangle the units or taxa of mineralogy from the idea of matter, which, still to this day, is conceived in terms of simple substance and pure form.

To return to the question of how the concepts crucial for the emergence of scientific mineralogy are telling of human and economic concerns, Werner's insistence on "outward characteristics" are an interesting case. They refuse to cut work practices, natural philosophy, or human sensual perception out of the equation. Yet, while humans began to disappear from the conceptualizations of scientific mineralogy, the latter still bear the mark of a distinct socioeconomic context. The abstractions of pure matter, which only emerged around Werner's time, clearly spoke to the logic of the production of money, and while the idea of pure matter superseded the perspectives of workers or dilettantes in mining and mineralogy, simple substance nevertheless has to be understood as an expression of the developed state of capitalism. This socioeconomic setting made mineralogy less explicitly entangled with human concerns on a conceptual level, but such concerns remained an implicit political economy inscribed into the modes of classification nevertheless.

Mineral Totemism as Thick Classification

Totemism as an expansive categorization system that incorporates both humans and nonhumans caught the attention and theoretical imaginations of European ethnographers around 1900. These researchers reported upon societies, located everywhere from North America to Oceania, that were ordered around a particular *thing*—a thing that exerted an inexplicable power over the people and their behavior. Both organic and, occasionally, inorganic objects could be elevated to this eminent position of a "totem." From the start, ethnographers investigated totemism as a curious mode of classification that placed humans, in an uncommon way, in the same system with plants, animals, and sometimes even minerals. One of the first famous researchers studying the phenomenon, the ethnographer James George Frazer, maintained: "As distinguished from a fetich, a totem is never an isolated individual, but always a class of objects, generally a species of animals or of plants, more rarely a class of inanimate natural objects, [and] very rarely a class of artificial objects."[23]

This class of nonhuman things served to bind human social ties. Totems reigned over sexuality, religious belief, consumption, war, and peace. The term for this comprehensive classification, "totem," was derived from an Anishinaabe (Ojibwe) word.[24] Questions about totemism appeared on the first pages of ethnographic questionnaires,[25] Indigenous travelers on steamboats traversing the Pacific were pressed for information about their "native lands," missionaries were cross-examined about their observations, academic journals published series on totemism, and, eventually, experts from the aforementioned Frazer to Sigmund Freud wrote monographs on the subject.

In 1900 the ethnographer W. H. R. Rivers observed how water, fire, and a bowl were raised to the status of a totem, in a small island between Australia and Papua New Guinea.[26] Rivers even met a local informant who claimed his personal totem, or *atna*, was "a large

23 James George Frazer, *Totemism*. Edinburgh: A & C Black, 1887, p. 2 (01-25-2008), https://archive.org/details/totemism00frazuoft
24 A frequently cited mention goes back to a fur trader and traveler: *John Long, Voyages and Travels of an Indian Interpreter and Trader Describing the Manners and Customs of the North American Indians; With an Account of the Posts Situated on the River Saint Laurence, Lake Ontario, & c.*, ed. Reuben Gold Thwaites. 1791; repr. Cleveland, OH: Arthur H. Clark, 1904 (09-28-2010), https://archive.org/details/cihm_36367
25 James Georges Frazer, *Questions on the Customs, Beliefs, and Languages of Savages*, 3rd ed. Cambridge: Cambridge University Press, 1916, pp. 13-14 (07-12-2018), https://hdl.handle.net/2027/uc2.ark:/13960/t1ng4qf38
26 William Halse Rivers, "Totemism in Polynesia and Melanesia," *Journal of the Royal Anthropological Institute of Great Britain and Ireland*, vol. 39 (Jan.-Jun., 1909), pp. 156-80, here p. 166. This information reportedly came from Rivers' missionary contacts Rev. W. J. Durrad and Rev. C. E. Fox. See William Halse Rivers, "The Terminology of Totemism," *Anthropos*, vol. 9, no. 3 (1914), pp. 640-46.

stone called Kalinga."[27] While this case of "mineral totemism" is rather rare, specialists estimate that around one-fourth of the recorded cases of totemism diverge from the more typical animal and plant totemism, and promote body parts, abstract numbers, or clouds to the status of ancestors.[28]

Societies structured according to this principle displayed a particular social order. People conceived of themselves as descendants of the totem, and the totem also organized relations between groups unconnected by kinship but held together by a shared strong relation to the chosen entity. These relations described a specific logic of care: even if people all came from the same village, those who belonged to different totem groups might be treated in a more negligent fashion.

What is more significant, here, to a history of the human–mineral relationship is that aspects and responsibilities assigned to the totem object also extended to members of the totem group. Thus, the totem could mandate certain behavior; for example, "the *wire* or water people may not drink the water of a certain bubbling pool; the members of the *tegmete* division may not eat food prepared in a bowl and the *ambumni* people may not walk on grass."[29] These restrictions are typical characteristics of totemism. Thus, the system not only proposes close, intimate ties between humans and nonhumans but also sets prohibitions and rules of usage. Humans are beholden to the totem and are required to abide by certain taboos. Grass or prey must be protected; walking and hunting are prohibited. A functionalist perspective readily interprets these taboos as resource management. While this remains a much-debated approach, what is uncontroversial is that the totem order impinges on social order and interhuman

---

27 Rivers, "Totemism in Polynesia and Melanesia," p. 160.

28 Philippe Descola, *Beyond Nature and Culture*, trans. Janet Lloyd. Chicago, IL: University of Chicago Press, 2013, p. 159. Fr. orig. *Par-delà nature et culture*. Paris: Gallimard, 2005. Descola refers to the comprehensive survey on totemism in Australia by A. P. Elkin from the 1930s, but also to more recent accounts: Carl Georg von Brandenstein, *Names and Substance of the Australian Subsection System*. Chicago, IL: University of Chicago Press, 1982; Carl G. von Brandenstein, "The Phoenix 'Totemism,'" *Anthropos*, vol. 67, no. 3/4 (1972), pp. 586–94; David Ludwig, "Indigenous and Scientific Kinds," *British Journal for the Philosophy of Science*, vol. 68, no. 1 (2017), pp. 187–212.

29 Rivers, "Totemism in Polynesia and Melanesia," p. 167.

relations in such a way that property and inheritance become organized along the same lines. That is, the same classificatory scheme serves as natural order, economic order, and social order in one. Resource economies, then, can be equated to totemism, in the sense that this classification determines usage of food and resources, determines obligations to care, and sometimes mandates the non-use of nature, which is understood as kin.

Although integrating all the rules governed by a totem object into one's life might seem overwhelming, the contemporary ethnographer John Comaroff compares totemism quite favorably to societies structured around ethnicity. In totemism, the various social ties are structurally similar and less integrated into a dominant whole, whereas "ethnicity has its origins in the asymmetric incorporation of structurally dissimilar groupings into a single political economy."[30] Comaroff contends that ruptured power situations and post-conflict environments could act as potential catalysts for the development of totemism. He is interested in social ties that depend on natural objects much more than on bloodlines. Yet the particular form of classification enacted by totemism does not depend on inherited bodily characteristics, and so it does not perpetuate inequality in the same way as classifications according to ethnicity do, which naturalize belonging according to physiological traits that differ from a mainstream society.

Today, totemism as a framework through which to rethink or even reformulate relations between humans and nature is having a surprising revival. The ontological turn in ethnography offers an interesting take on older formulations of totemic taxonomies. For example, totemism looms large in Philippe Descola's four ontologies (animism, totemism, analogism, and naturalism) and also speaks to Eduardo Viveiros de Castro's perspectivism, which spells out divergent modes of classification from the Amazonia.[31] Such contemporary theory makes totemism (or animism) sound tantalizingly ecological.

30 John Comaroff, "Of Totemism and Ethnicity: Consciousness, Practice and the Signs of Inequality," *Journal of Anthropology*, vol. 52, no. 3/4 (1987), pp. 301–23.
31 Eduardo Viveiros de Castro, "Cosmological Perspectivism in Amazonia and Elsewhere," in Castro, *Four Lectures Given in the Department of Social Anthropology, University of Cambridge, February–March 1988.* Manchester: HAU Books. Masterclass Series 1, Network of Ethnographic Theory (2021), pp. 45–168.

This recent interest is all the more surprising given the role totemism played in supporting a surprisingly comprehensive list of ethnographic methodologies, from the pitfalls of evolutionary theory to the racism of *Kulturkreislehre* (cultural field school), and from narrow functionalism to the very heart of structuralism. Amid attempts to arrive at multispecies perspectives and non-Northern sustainability, totemism appears to be proving itself an unwieldy tool, which necessitates an analysis of ontological dimensions more broadly.

Notably—and maybe misleadingly—all this modern theorizing about this classificatory scheme points toward one latent promise: Could this differently ordered world offer an ecological advantage? It's not easy to answer this question in the affirmative. Simply putting the romanticized elements of kindness, kinship, relationality, and ecological connectedness front and center does not help much. Nevertheless, considering how the conceptualization of matter prohibits or encourages behaviors and how it legitimizes and predetermines resource allocation is a timely and important investigation.

Instead of reifying this mode of human–nonhuman classification as ontology or cosmology, we could instead perceive it as a discourse, a particular way of asking questions and making sense of things, which is a way to perceive of the world, not its inherent structure.

Remarkably, this attempt to arrive at epistemological perspectives is what Descola seems to reject in perhaps his most famous predecessor, Claude Lévi-Strauss. For the latter, the totem animal, plant, or mineral is a resource for organizing differences. Totemic classification firstly allows for an intellectual order, and only subsequently are social groupings and social behaviors affected. For Lévi-Strauss, two images stabilize each other: one related to natural difference, the other to social belonging and responsibility. They occur on one inseparable plane of existence, governing nature as well as society with the same kind of law.[32] Descola rejects Lévi-Strauss' key fascination with this power of categorization, which he deems intellectualist, abstract, and

32 Claude Lévi-Strauss, "Der Totemismus von Innen," in *Das Ende des Totemismus*. Frankfurt am Main: Suhrkamp, 1962, pp. 120–35. Fr. orig. *Le totèmisme aujourd'hui*. Paris: Presses Universitaires de France, 1962. Engl. trans. Rodney Needham, *Totemism*. Boston, MA: Beacon Press, 1964. For an earlier epistemological approach see Richard Thurnwald, "Die Psychologie des Totemismus," *Anthropos*, vol. 12/13 (1917/18), pp. 1094–113.

infested with dichotomies. Instead of explaining social hierarchies as consequences of the capacity to group and regroup nature into a taxonomic logic, Descola's world seems to structure itself more easily as a welcoming plenitude. There is no need to find cognitive dichotomies and sustain them in the outer world, to give it the structure of a totem and totem groups. Social discontinuities are not learned and transferred from conceptual ones, to Descola's mind. He stresses the divergent anthropology of nature, which does not split visible matter into substantial and incidental, into true idea and negligible occurrence, humans and minerals, but rather departs from mighty streams of sameness, underlying matter, and "individuated organisms" alike.

Totemism is successful in dispensing with separate taxa. It encourages identification with natural elements—sometimes rocks—as totem groups while excommunicating fellow humans into another order. Under totemism, humans do not emerge from Adam's rib. We all branch off from a first being, which remains part of the new entity. Human relation to nonhumans can be very deep in these schemes. The world is multiple flows of admixtures, a process that not even time can keep in check in the way we would expect descent or generation to work. The relation of the whole totem group exists outside the present and continues to resonate with processes that originated at the beginning of history. One key example of totemic conception is the "dreamtime" of Aboriginals in what is currently called Australia. In dreamtime, the beginning of mankind is actively felt in the present. Dreamtime coexists. It is by no means an epoch of the past, a time to remember, and neither a possible future.[33] The ties binding people and landscapes into units over time are vivid and substantial lived realities, not abstract taxonomies. What Descola terms "relation" is more unrelenting than a mere psychoanalytical emotion felt toward an object. It is richer and less voluntary than the insights that arise in a shared classificatory order.[34]

While Descola describes naturalism—the typical ontology of the Global North and a counterpart to totemism—in critical terms, and lists it as one of the fourfold modes of being, he writes much less about actual nature than one might expect. Naturalism divides humans into a

33 Descola, *Beyond Nature and Culture*, pp. 146–47.
34 Ibid., pp. 112–15.

physical and spiritual existence: half angel, half animal.[35] While Descola's strongest move seems to be the distance he claims to old-world, self-evident, intellectualist concepts, there is probably no better reading than Descola to experience the profound difficulties of seeing beyond them—to reconceptualize nature, to leave the world of simple substance and to follow other modes of human–mineral adhesion.

Outlook on the Normative Elements of Anthropocene Classification

Compared to the history of totemism, the twofold disarray that technofossils cause within mineral classification may seem faint, tame, or even narrow. Nevertheless, the suggestion of this text is to value them as a conceptual revolution judged against the history of mineralogical taxa. Given that technofossils are a still novel but widespread physical phenomenon, geological conceptualization has to eventually adapt to the new realities produced on Earth. The small provocation posed by this conceptual innovation in geology is thus still a part of the natural sciences, but it cannot avoid stirring up disarray. Technofossils developed in analogy with petrified forms of life, namely fossils. Even trace fossils fit into categories set according to mineral taxonomies, yet their names pay witness to some activity of living beings, albeit in a way much smaller than the agency claimed by humans, who have become a geological force. Fossils are categorized as, for example, *domichnia* (dwelling structures), *repichnia* (surface traces of creeping and crawling), or *fugichnia* (escape structures).[36] Technofossils were modeled on these taxa and incorporate the agency of living organisms. They provoke the clean-cut order of simple substance, in that technofossils always imply human activities and prohibit stripping the classificatory orders of any social meaning. As was shown during our dip into the history of mineral classification, simple substance—neutral as it may seem on the surface—can be read as an expression of a particular economic activity or desire. Admittedly, these economic meanings remain implicit and surface only in historical perspectives.

---

35 Ibid., pp. 170–73.
36 Adolf Seilacher, "Sedimentological Classification and Nomenclature of Trace Fossils," *Sedimentology*, vol. 3, no. 3 (1964), pp. 253–56.

Technofossils do, of course, have an industrial signature and can by no means be equated with totemic classifications, which are typically found in nonindustrialized societies. What is evident, though, is that both of these modes of classification make the human–mineral relation explicit. While in totemism classes of objects exert a power that structures behavior toward other humans and nonhumans alike, the present upsurge of new matter embodies a different relation. Neutral and relationless human agency is what produced and amassed this new layer of technofossils around the globe, which will never be quite natural again. Thus, all three modes of classification—Enlightenment-era mineral classification, totemic classification, and the new taxa of the Anthropocene—speak to a typical economic structure. The fate of society, or at least some element of economic significance, is pulled into the wake of the mineral classificatory system. All three modes of classification imply human agency and mirror economic ways of conceptualizing minerals and organizing resource access and protection. Even seemingly neutral mineral classifications justify certain uses of matter. What is newly introduced by technofossils, however, is that the normative element encapsulated by this notion seems to explicitly suggest a misuse of resources. It could be argued that this normative element makes mineralogy more human compared against the history of mineral classification, and due to the new physical dependence of technofossils on human production it may even be considered a step toward the strong links of humans, though in reversed order compared to totemism. This makes thinking with technofossils a very promising conceptual revolution.

1786.

Pom: 3.

Who Defines the Anthropocene?

Victor Galaz: The Anthropocene concerns literally everybody. Yet the AWG is made up of only a tiny fraction of the planet's population. Is the question of representation an issue among the members of the AWG?

Simon Turner: You are absolutely right. The Anthropocene is a planetary-scale subject, so how come only twenty-two people make the decisions? Twenty or so people of any group on a planetary decision seems an incredibly small number, but it is to do with the protocols of defining geologic time scales. From the start, we have tried to make this process exactly the same as with any other geological designation of age, stage, or epoch. We go through the same process as if the Anthropocene was 10,000 or 100 million years ago.

VG: But is going through the same process a good thing? The context of the Anthropocene is evidently different to previous geological time units, is it not?

ST: Stratigraphy is a very conservative subject. It attracts people with a lot of thought, consideration, time, and effort, demanding they ask: "Do we call this a new time period, or do we call it an epoch?" It is actually sort of pedantic—people get upset in geology if it's 2.4 million years instead of 2.45, because scientists quite rightly like precision. But the amount of people that get upset about that is not usually massive. Within a broader scientific community, such decisions have implications, but up till now, the implications have not gone far beyond the field of scientific geology. But those are the rules that we have to abide by.

Politically, in our relationship to the ICS [International Commission on Stratigraphy], this is a very interesting question. Do we just turn up and say, "Hey, we were thinking of changing the rules"? I don't think that would go down too well. But who knows, maybe they would say, "you're right; you're absolutely right. This is a thing too big for such a small working group. This is fine for any other time unit but not for the Anthropocene, we should do it differently." That would be interesting!

VG: That would mean an institutional innovation, right? Think of the IPCC [Intergovernmental Panel on Climate Change] as an institu-

tional science-policy innovation. It didn't exist before. At some point, someone said that actually, the way we're currently doing science and interacting with policy-making is not working. The way we've set up collaborations with society and policy-makers is not working. We need to innovate and create something different.

ST: But the AWG came from a very different place—it came from the field of stratigraphy. We realized that the Anthropocene had already been recognized in Earth system science and wondered if it would be significant enough to leave a geological mark. And because it originated from there, it came with all the instructions around how to define a stratigraphic period, how that mechanism works, the institutions that are required to justify that, and so on.

But to our credit, from the start, the group was not made up solely of geologists. The SQS [Subcommission on Quaternary Stratigraphy] suggested we "assemble a group of people with expert knowledge of that time period." So the AWG did actually assemble quite diverse expertise from the fields of archaeology, history, law, and Earth system science, which is very unusual for a geological group. Also, we have always worked on promoting a broader understanding. For instance, the cooperation with HKW has enabled us to have these bigger discussions with artists, historians, et cetera.

VG: Do not get me wrong, I feel nothing but respect for the work conducted by the AWG. And I think few people from outside academia realize how much free time scientists spend on groups like this, out of a sheer passion for the science—and for society.

The above conversation is an edited excerpt from "Exchange on Collaboration and Complexity," a discussion held on May 21, 2022 at HKW in Berlin during the event *Unearthing the Present*.

On Human and Coral Structures

Kristine DeLong: Coral is a simple animal, elegant in some ways. Its skeleton is not inside the coral animal, it's on the outside, and the coral lives on top of its skeleton. The skeleton grows kind of like a skyscraper. It'll put in a floor, then walls, lift itself up, and then put in another floor, continuously through time.

As it's making the skeleton, the chemistry of the water becomes part of that skeletal framework that it is building, so if we look at different elements in the skeleton—like strontium and calcium—we can calculate what the temperature was when that particular floor and ceiling was put in. And because we have these quite precise chronologies that the coral leaves us, we can go back and say, for example, "in 1950, the water temperature was this temperature."

Nigel Clark: I am in no way an expert on corals, but I love the provocation that they offer us. They offer this sensorial universe of a very, very different creature from us. Its senses are different, its body plan is different, the organisms that it collaborates with are very different from us. But what really strikes me is the number of things that we have in common with corals.

Kristine compared coral reefs to skyscrapers. And this metaphor really works wonders for me. It's more than a metaphor in some ways, because coral reefs are indeed literally biogenic constructions. And there are not many other organisms I can think of, apart from us—bipedal hominins—who build biogenic structures that are hundreds or thousands of feet high. What have we got in common? Why do we both work the planet in three dimensions? I was going to say that we both work the planet so incredibly well, but really, whilst that's true of the corals, I'm not so sure about us.

It strikes me that one of the things we have in common is that humans emerged, as far as we know, in the East African Rift Valley, a very tectonically active area with lots of volcanoes. And we learned to negotiate in three dimensions in this kind of rift landscape. Corals also cluster around volcanoes and plate junctures, and they reflect past continental plate collisions that created shallow seas. So, what we've got in common is that we both negotiate an Earth that is highly tectonically active. And we somehow, both us and the corals, learn to respond to this very, very changeable Earth by creating gigantic structures.

Some of the earliest kinds of structures that humans inhabited were limestone caves. We learned to inhabit mineral structures by going into limestone caves that were possibly constructed from corals. Later on, we started constructing cities, we started "mineralizing" in a very similar way to how corals seem to mineralize, we started creating a kind of exoskeleton for ourselves. Think of the early city walls, those skeletons in the sand that grew and grew. One of the reasons they grew was in response to floods from alluvial plains. City walls were not just keeping enemies and wild animals out, they were also keeping out floods. They, in themselves, were a kind of reaction to changeable, variable, unpredictable weather.

Jens Zinke: This is a beautiful summary of the similarities between corals and humans. Coral reefs are indeed, in a way, underwater cities. It is not very different from how we live. We organize ourselves; the corals organize themselves. Some corals don't like the others, so they even start fighting each other with chemicals. Well, let's hope that we will never do that again!

There are these kinds of interactions between different organisms everywhere within the reef; there are "helpers" and "cleaners"—it is a fully functioning ecosystem, not very different from ours. One difference could be that corals cannot talk. But maybe they can, we just haven't discovered it yet.

Kat Austen: One of the beautiful things about coral reefs is how lush and varied they are and how they are a home for so many different species. And indeed, humans do create that, sometimes intentionally, which is rather beautiful. This reminds me of the initiative "Animal Estates" by the artist Fritz Haeg. The project consisted of a series of experiments in the design of urban environments for nonhuman others, ones that made sure the urban environment would be welcoming to other species.

KDL: There are also very literal connections between the construction of coral reefs and human dwellings. Many of the island communities in the Caribbean use coral rock for their building materials. In the Cayman Islands, you find fences all over the place that are actually coral rock. If you go down to Miami, most of their government buildings consist of Key Largo limestone, which is all former coral reef that is over 130,000 years old.

NC: It is also very interesting how much of the language that we use to describe the dynamics of coral ecology echoes the challenge of human migration. Recent research has shown that some young coral in warming tropical oceans are able to use currents to move into cooler, sub-tropical waters and establish themselves there. But then you think about how incredibly difficult it is for a human to migrate, given the same pressures from climate change, and the time it takes for different human communities to establish themselves in a new location. This is actually both tragic and ironic given the fact that there's evidence that early humans migrated across the surface of the planet, partly in response to climate change but also following rugged topography. They followed the topography they were familiar with: the rifted and rugged and sometimes tectonically active landscapes. So, like corals, early humans followed volcanoes and rifts.

But what are we doing now? We are building walls and putting up barbed wire precisely to stop humans doing exactly the same thing that we and other organisms do when adapting and responding to climate change.

The above exchange is an edited excerpt from "Conversations Beyond the Human," a discussion held on May 20, 2022 at HKW, Berlin, during the event *Unearthing the Present*.

C.ᵘ No: 72.

Or: 21.

Ice Core Temporalities

Susan Schuppli: Ice is something that on the one hand is very familiar to most people, while, on the other, it is extraordinary. In particular, it can provide us with the highest resolution datasets of climatic changes and with insight into the ancient atmospheric histories of the Earth.

The other day, we looked at the ice-slice samples you brought and discovered that the air bubbles in the ice are actually younger in age than the ice that had captured and archived that air. I think these double temporalities are quite fascinating for those of us who are involved in the humanities.

Liz, how do you engage with these multiple temporalities that are archived in ice, and in what ways do you pull apart these two different kinds of timescales?

Liz Thomas: Well, think about the snow falling: the fresh snow is all nice and fluffy. And when you collect all that fresh, fluffy snow and put it together to make a snowball to throw at someone, you want it to be really hard. To do that you compact it, and it changes the density of the snow. And within the ice, these bubbles are formed.

But what happens in the ice sheet is that you've got this snow falling year after year. It gradually builds up and up and up. The snow at the bottom is getting buried, compressed, and becoming denser. This can take quite a long time, depending on where you are in Antarctica. It depends on the temperature, how much snowfall you get per year, and how quickly that densification process occurs.

This means that you've got potentially anything from between maybe fifty to a hundred meters where there's still this snow that we call firn. Here in the firn, the bubbles haven't actually closed off and formed yet, they're only starting to form. They have little channels in between them, so that the air can move not only along the ice sheet but also up and down, and all the while can also go through changes based on the weather conditions at the surface.

This means that there's this whole section of the ice where the air can still be moving around. And it is only at the point where the ice reaches a critical density and the bubbles are closed off that the air becomes trapped. And that's how you can then find that you've got quite modern air, potentially fifty meters down.

SSCH: This brings me to the question of the ice-sheet dynamics. I'm guessing that they're similar to the dynamics of tectonic plates, where an older layer can eventually end up on top of a younger layer. How do you deal with glacial folding when trying to ascertain the temporality of an ice core?

LT: This does pose a problem. Especially in some of the very deep Greenland ice cores, the ice on the very bottom can be twisted and folded, which can make the interpretation of it quite tricky. The way we address this is by doing a detailed geophysical survey of the sites we intend to take a core sample from. In general, we try to aim for drilling locations that are as stable as possible. A glacier is moving, like a river, albeit very slowly. Therefore, we avoid drilling in the ice flow itself, where the glacier might have been moving for thousands of years. But the central area at the very top of a glacier, the so-called ice divide, might be comparably stable. Radar surveys also allow us to "see" through the ice all the way to the bottom. On the radar, if there is any kind of disturbance in the ice, it looks a bit like spaghetti. In this case, we would probably choose not to drill, so it does come down to being selective and really understanding your site.

Reconstructions of the Unseen

Susan Schuppli: I have always wondered why we need to continue ice-core drilling when the cores we have already have been very persuasive by giving us this clear "hockey stick" graph. These cores have already made the case that there has been anthropogenic climate change, especially in terms of atmospheric greenhouse gasses. But it's become clear to me that ice cores can be used for many other purposes. What are some of the other things that you're looking for in these ice samples?

Liz Thomas: One of the things that have been very persuasive, as you said, is this "hockey stick" graph, informing us about the climate, and particularly about changes in greenhouse gasses. But those are only the kind of headline stories. Actually, we have still only scratched the surface of what other information we can find contained within the ice.

Just to give you a few examples: one of the other things we can do with ice cores is to reconstruct past sea-ice levels. This is because of the chemicals produced by the marine algae that live in the sea-ice zone. Antarctica is a pretty inhospitable place, but there are a few boundary areas where there is quite a lot of productivity, for instance around the edges of the sea ice.

Those little algae emit chemicals that are then transported to the ice where we can detect them. They can also be lifted off the surface of the ocean by strong winds and can then actually be transported onto the ice. They're carrying with them all this information about how productive they were. And that tells you something about how far the sea ice extended in time. You see, sea ice is a really good example of something we can reconstruct. And sea-ice data is also suited to bringing in other records beyond the ice cores.

It wasn't until 1957–58 (the International Geophysical Year, IGY) that the first climate observations started to be put up. Prior to that, however, Antarctica had been visited very actively for whaling—it was a prime site for hunting down there. First it was the seals, then whales. A huge industry sprung up, and the whale populations were almost obliterated.

Though this in itself is another example of quite uncomfortable information coming from an uncomfortable time, now some valuable data is emerging, because these hunters kept all the ships' logs.

We've actually got a record of where ships were going to hunt the whales. And we can then pull back this information to help us do these longer reconstructions as well, which has been really, really valuable.

For example, South Georgia is a beautiful, absolutely stunning little island. Famously, it was the island Ernest Shackleton and two of his men crossed in their rescue attempt. He was able to return and rescue the other men specifically because he was aiming for the whaling stations. It was the huge whaling operation and enormous population of whalers that enabled him to mount the rescue.

At the time, there were around 2,000 people living on this tiny island, constantly bringing in whales, leaving them to die on the shore, and then boiling their carcasses in cauldrons to extract the blubber to produce whale oil. Then you think of all the contaminants from the fires for this boiling and what the process would be emitting while the carcasses are bubbling away. All that emission would be going straight into the ice, straight into the ocean. And potentially there is still a lot there.

SSCH: I was reading something about actually being able to read past wind and storminess in an ice core. I found that completely baffling. How is it possible to read wind in solid material?

LT: This is another example of that reconstruction of the unseen. You cannot see winds, you can only see what gets around moved by the winds. There are a few different ways you can do this. For example, you can melt ice on a hot plate and count the number of particles in the ice. Normally we use this method to see how much dust is in the ice. You'd count the different dust particles and put them into different bins based on their size.

The dust is believed to come from the Southern Hemisphere landmasses, predominantly Patagonia. Smaller amounts are probably from Australia and lesser amounts from South Africa. When you look at an ice core, you can see—well, if it's particularly dusty, we can assume that potentially the winds were very strong because they could bring more dust from farther away to the site.

The only problem with this is that changes in dust are related to the local conditions. For example, we know that Patagonia got drier during the twentieth century. It can be quite difficult, therefore, to disentangle what we're actually picking up as a big increase in dust and what is actually in the ice core. Is it telling you about Patagonia getting drier, or is it telling you about the winds that brought them being stronger?

One thing we've been trying to do is to look at the marine diatoms in the ocean. These small unicellular algae are ubiquitous, you get diatoms and algae everywhere. And they are living in the surface microlayer of the ocean. But under strong wind conditions, as the waves bring up the diatoms, they get lifted up into the atmosphere. Then the winds transport them and they get deposited onto the ice sheet. And I don't mean transported to the ice sheet just nearby, they get transported thousands of kilometers from where they were originally happily living on the surface of the ocean.

We are filtering the algae while we are collecting the meltwater. We identify these diatoms by looking at them under a microscope. And a key wind-related thing about them is the sheer number of diatoms. So, when we get a slide that's absolutely jam packed with these diatoms, we can assume that there had to have been very strong winds.

The other thing we can do is actually identify what species they are. We can tell you exactly where in the ocean we would expect to find these particular diatoms. So that's how we know that actually, for a lot of these sites, particularly at this Palmer Station in the north of Antarctica, the diatoms found there didn't live near the sea ice or near the Antarctic, they lived right out in the South Pacific. And it is these diatom species that have been transported. It is looking like a really good proxy for winds.

We know this because we can compare it with the instrumental data. We have this beautiful period of overlap, where we can compare it with the winds. And the other advantage over some of the more traditional wind proxies such as dust is that the diatom species and composition in the Southern Ocean won't have changed that much over hundreds, even thousands of years.

SSCH: That is really amazing. And because there are two layers of snowfall each year in the North and South Poles, there is a biannual record that provides an extraordinarily high resolution. I guess, therefore, that we would know a period of windiness for a six-month sort of window, is that correct?

LT: With the ice that we're looking at, every snowfall event is recorded. This means that in a lot of these sites we can go back to the individual storm events and know exactly what season it was, down to almost the week, the day.

SSCH: What is the maximum number of events that you've encountered in an ice core?

LT: In the ice cores we are working on currently from the Antarctic Peninsula, the majority of the air masses are actually coming across the Amundsen-Bellingshausen Seas. And this is one of the most variable weather systems on the planet. It has a huge amount of these very deep, low pressure storms that come through—probably in the order of a hundred continual, big, deep-storm systems that deposit loads and loads of snow on the Antarctic Peninsula and everything else that the snow brings with it.

SSCH: Are you saying that in the period of a year, you could actually discern maybe a hundred different storms out of the ice-core record?

LT: Yes, though that would be a very laborious approach, the data is there. Mostly we just look at either the seasonal events or at the annual record, because we take it back hundreds of years.

The above conversations are edited transcripts from "Exchange on Melting Narrations," a discussion held on May 22, 2022 at HKW in Berlin during the event *Unearthing the Present*.

Nᵒ 1294

Carbon Aesthetics Group*

Whale Falls, Carbon Sinks: Aesthetics and the Anthropocene

Valuing Whales

In 2019, economists at the International Monetary Fund (IMF) calculated that whales, when considered a carbon sink, are an ecosystem service which may be worth millions of dollars per whale.[1] A few years earlier, a number of new studies showed the harmful effects of noise pollution on these marine animals.[2] Each approach sees the whale through a different set of material relations in which it is embedded: one is economic and calculative, and the other sensory, aesthetic, and affective. For the stratigraphic identification of the Anthropocene, the chemical signatures of large-scale processes play a determining role. The protocols of stratigraphic classification, which make particular chemical histories not only visible but authoritative in categorical designation, prompted our consideration of which socio-material relations with chemicals are rendered visible and operational through various processes of the techno-sciences. These processes have been called the everyday politics of aesthetics, which operate beyond the divide between the sensuous and the intelligible.[3] They have driven

* Karolina Sobecka, Desiree Foerster, Myriel Milićević, Alexandra Toland, and Clemens Winkler

1 Ralph Chami, Thomas Cosimano, Connel Fullenkamp, and Sena Oztosun,"Nature's Solution to Climate Change: A strategy to protect whales can limit greenhouse gases and global warming," *Finance & Development* (December 2019), pp. 34–38.

2 Marcos R. Rossi-Santos, "Oil Industry and Noise Pollution in the Humpback Whale (*Megaptera novaeangliae*) Soundscape Ecology of the Southwestern Atlantic Breeding Ground," *Journal of Coastal Research*, vol. 31, no. 1 (2015), pp. 184–95; Nicola Jones, "Ocean uproar: saving marine life from a barrage of noise," *Nature* vol. 568, no. 7751 (2019), pp. 158–61.

3 Margus Vihalem, "Everyday aesthetics and Jacques Rancière: Reconfiguring the common field of aesthetics and politics," *Journal of Aesthetics & Culture,* vol. 10, no. 1 (2018), https://doi.org/10.1080/2000 4214.2018.1506209. All online references in this essay were last accessed in October 2022.

the knowledge practices and sensibilities made manifest in everyday products and lifestyles—from scrimshaw snuffboxes to Twitter "fail whales"—that span centuries of consumer behavior and have an impact not only on the individual psyche but on social relationships driven by what Gernot Böhme describes as "aesthetic capitalism."[4] Understanding these processes as generative of a certain "regime of visibility" or a certain aesthetic allows us to consider how they shape environmental realities. It invites us to ask, as Nicholas Shapiro and Eben Kirksey write, what the infrastructures are that generate environmental constructs, that can perpetuate environmentally embedded violence, and where the risks of reproducing them or the opportunities of countering them lie.[5] When our chemical coexistences are always mediated, what role can aesthetics in the sense of aisthesis, or sense-perception, play in aesthetics as a mode of performing politics? What role can it play in influencing value assessments of more-than-human lives such as our mammalian relatives in oceans deep? As Birgit Schneider suggests, perhaps a path to action must be led via aesthetics.[6]

The point of departure for this contribution are studies that consider whale bodies as an environmental archive, containing many of the material markers that signal the onset of the proposed new epoch of the Anthropocene and in particular chemical markers of fossil fuel combustion. Given that at least fifteen cetaceans species are listed by the International Union for Conservation of Nature (IUCN) as critically endangered, over the future of whales looms another possible marker of Anthropocene—species extinction. But it's not only the presence of material markers that led the group to focus on whales. Choosing whales centers a nonhuman being, and invites a closer consideration of exactly *how* whales and their material relations are invisible. What registers of perceptions and sensitivities would enable

4   Gernot Böhme, *Critique of Aesthetic Capitalism*. Milan: Mimesis International, 2017.
5   Nicholas Shapiro and Eben Kirksey, "Chemo-Ethnography: An Introduction," *Cultural Anthropology*, vol. 32, no. 4 (2017), pp. 481–93.
6   Birgit Schneider, "Klima – Daten – Kunst: Künstlerische Aneignungen atmosphärischer Forschung," *Informatik Spektrum* vol. 44, no. 1 (2021), pp. 50–56.

an understanding of material relations in which it and us are embedded together in a way that would foreground and reorient unrecog-
nized connections, concerns, and values?

Sensing Whaling

Following the different interconnections that have been shown between whales and their ecosystems, we inevitably retrace the paths
forged by the human knowledge practices that produced and privileged a particular mode of relating, the very mode which set in motion
many of the "accelerators" of Anthropocene processes. The human
transformation of environments for economic productivity became
disastrously linked with whale species through the emergence of industrial whaling in the seventeenth century. Even though commercial
whaling was banned in 1986 by the International Whaling Commission (IWC), whale populations have never fully recovered. While the
practice of whaling is documented to be at least 8,000 years old, the
large-scale "harvesting" or "mining" of whale bodies for fuel is, like
so many other aspects of the Anthropocene, inherently linked to processes of imperial colonialization and industrialization. Whale oil
and spermaceti, the energy sources derived from whale tissues, were
used primarily for lighting, for producing an ambiance, an environmental condition which enabled the production of knowledge and
subsequent industrial development.[7] Burning of whale oil advanced
humanity on the closely entangled paths of illumination, productivity,
carbon fuels, and combustion.

    After spermaceti was first used to make candles in the mid-eigh-
teenth century, the lucrative potential of whale-based illumination
products was quickly recognized and spurred a long-lasting, immense-
ly profitable, and politically-influential whaling industry. Only after
more than a hundred years of dominance did whale oil start to be re-
placed in the late nineteenth century by kerosene, or "coal oil," invented in 1846 by the Nova Scotian physician and geologist Abraham
Gesner. The subsequent energy transition happened fast, propelled by

---

7   Voltaire Foundation, *Voltaire Foundation Blog*, "Tag Archives: Whale Oil,"
    https://voltairefoundation.wordpress.com/tag/whale-oil/

a number of factors. Whaling was in decline, as whales had already been hunted to near-extinction and hunting vessels had to make longer and more risky voyages to find them. Whale products shot up in price, while petroleum, one of the sources for kerosene, was cheap to produce. In addition, petroleum was becoming a material feedstock which found its way into a new consumer society in a variety of uses other than lighting. More complex refineries invented new technologies and products that served the new consumers.[8] Ultimately, in a toss-up between kerosene and whale oil light, the quality of kerosene light became decisive for its adoption: kerosene burned cleaner and more brightly. Even lighthouse keepers, loyal holdouts to the whale oil industry, had to convert to kerosene when ship captains complained about the inferiority of the whale-oil lighthouses.[9] As European seafarers favored the light that shone most brightly, whale populations were inadvertently enabled to survive, even though their numbers remain critically low.

Lighthouses, these remote points of light signaling the furthest reaches of European exploits, guided the vessels and the processes of globalization in the nineteenth century, beguiling voyages of colonial expansion, trade, and migration. As the paths for exploration, extraction, and production were revealed and forged by literal and metaphorical processes of illumination—where artificial light and knowledge practices made the Earth ever more visible and better articulated as a resource—vision and its way of seeing carbon were aligned to make the Earth's matter productive. With the availability of artificial light powered by carbon fuels, vision was further privileged as a mode of sensing; in turn, vision privileged carbon as an object of sensing, sustaining a feedback loop of visibility, power, and aesthetics.

The same blubber that whales were hunted for, today proves valuable for creating another kind of visibility, extracting Anthropocene-marker relevant data. The analysis of the Carbon-13 (-13C) isotope, in whales' ear plugs and fatty tissues, shows with a six-month

---

8   Mark Foster, "New Bedford—Whale Oil Refining Capital," *IA: The Journal of the Society for Industrial Archeology*, vol. 40, no. 1 (2014), pp. 51–70.

9   Ed Buts, "The cautionary tale of whale oil," *The Globe and Mail* (October 4, 2019), https://www.theglobeandmail.com/opinion/article-the-cautionary-tale-of-whale-oil/

resolution the so-called Suess effect, the chemical signature of increasing levels of anthropogenic fossil fuel combustion over time.[10] Analyzing the tissues of multiple whales reveals a long-time series that crosses the potential threshold of the Anthropocene, proposed to be around the 1950s. Whale bodies are high temporal resolution (OCT) archives of other chemical histories as well. Fatty tissues act as natural sinks for lipophilic compounds, such as historic-use pesticides, polychlorinated biphenyls (PCBs), polybrominated diphenyl ethers (PBDEs), methylmercury, or hormones. The time-series of these chemical signals have been used to reconstruct past ecosystems, and histories of how whale individuals and populations responded to the environmental stress factors and other impacts caused by anthropogenic activities—including the whaling industry, noise pollution, war, transport, and the leisure industry, and more recently, plastic pollution and ocean acidification.[11] We argue that those chemical signatures can serve not only as an analytical tool, but also as a mode of relating—to whales, and more broadly to other entities and processes of the Anthropocene, if we make sense of them through affect and aesthetics. What might it mean to become sensitive to the broader range of material relations in which whales are embedded, rather than seeing only the relations which continue to cast the whales, along with other creatures and environments, as a resource?

We are inspired by the aesthetics that emerge from marine ecologists' descriptions of the horizontal and vertical translocation of nutrients in the oceans by whales as "whale pumps," "whale conveyor

10 Farzaneh Mansouri, Zach Winfield, Danielle D. Crain, and Brooke Morris, "Evidence of multi-decadal behavior and ecosystem-level changes revealed by reconstructed lifetime stable isotope profiles of baleen whale earplugs," *Science of The Total Environment*, vol. 757, no. 4 (2021), https://doi.org/10.1016/j.scitotenv.2020.143985

11 See Stephen J. Trumble, Stephanie A. Norman, Danielle D. Crain, et al., "Baleen whale cortisol levels reveal a physiological response to 20th century whaling," *Nature Communications*, vol. 9, no. 1 (2018), pp. 1–8; and Fletcher M. J. Mingramm, Tamara Keeley, Deanne J. Whitworth, and Rebecca A. Dunlop, "Blubber cortisol levels in humpback whales (*Megaptera novaeangliae*): A measure of physiological stress without effects from sampling," *General and Comparative Endocrinology*, vol. 291 (May 2020), https://doi.org/10.1016/j.ygcen.2020.113436

belts," and "whale falls."[12] The throngs of phytoplankton, thriving on whale excrement deposited at the ocean's surface, or the benthic invertebrates, feeding on dead whale bodies fallen to the ocean floor, can hardly be described as "consumers." Economic metaphors fall far short of the richness of mutualistic entanglements within networks of biogeochemical relationships and flows. Scientists react with a lack of words and expressions of wonder to the ecological complexity revealed in events such as "whale falls" (sinking of a whale carcass that provides a sudden, concentrated food source and a bonanza for organisms in the deep sea).[13] Witnessing scientists' grasping at strange metaphors to describe it makes it very clear that there's always more going on than what can be comprehended with reason and language, and certainly more than what can be captured in abstractions of markets and quantified exchanges. Whale falls rip holes in such constructed realities by presenting an image of nature that transcends any hard lines humans draw between species, generations, life and nonlife, or producers and consumers. As such, whale falls present an opportunity to invent a new language, new speculations and stories, to ask new questions.

Artistic Practices

The large-scale material transformations discernible in geological and ecological data that might come to mark the start of the Anthropocene also make the Anthropocene feel, smell, taste, and look a certain way. In other words, material impacts, even at a planetary scale, are discernible through aesthetic registers. To make sense of the Anthropocene by means of these registers would call for a mode of encountering the world through the dimension of affect rather than cognition, experience rather than representation, sense rather than significance. For example, we can ask, if perhaps speculatively: how

12 Joe Roman, James A. Estes, Lyne Morissette, et al., "Whales as marine ecosystem engineers," *Frontiers in Ecology and the Environment,* vol. 12, no. 7 (2014), pp. 377–85.

13 US Department of Commerce, National Oceanic and Atmospheric Administration, "What Is a Whale Fall?" (February 26, 2021), https://oceanservice.noaa.gov/facts/whale-fall.html

does a marine mammal experience ocean acidification? Is it felt in the hunger, malnutrition, or stress related to the disappearance of their food sources? Can the softening of shells and skeletons of whale prey be detected through mouthfeel (or for our purposes here, baleenfeel)? We know whale song can become indiscernible across long distances in waters loud with human noise, but does it also sound different in a medium chemically altered by excess carbon dioxide, microplastics, and the further impacts of these changes in the plant, animal, and microbial community? A sperm whale's clicking song at 180 decibels can be part of a lively communication with other whales near or far, a way to "see" bait in dark ocean depths, or to scan human free divers. Considering that whales probably discern and enjoy acoustic qualities of their songs and vocalizations, which songs do they experience as most enjoyable or beautiful?

On the other hand, what kind of ambience did the spermaceti candles produce for the humans that used and made them? Were the whale-oil-fueled lighthouses visible to whales migrating along the coast? What did London smell like when its streetlights were filled with whale oil, processed from tissues so visibly imprinted with stress chemicals?

By asking such speculative questions, artistic approaches let us publicly stage the mechanisms that distinguish which specific dimensions, out of the breadth and complexity of environments, are deemed recordable, classifiable, or worthy of inquiry. What is understood as relevant and what is not included in or removed from the discourse because it is considered unimportant? Or in other words, which questions are cast as silly or fantastical—and why?

Artistic practices serve to frame the much-cited phenomenological gap between Anthropocene processes such as human-induced species extinction or global warming, and what we are able to experience in our everyday lives. To go a step further, artistic practices often aim to "operate beyond the divide between the sensuous and the intelligible," on the interdependent relationship between what can be experienced and what can be thought. Whatever London might have smelled like with its whale-oil lamps, people would not relate the smell to the stress or extinction of whales because these simply were not concepts that eighteenth-century Londoners had. Given today's relatively new awareness of human impacts on the environment, the same smell might be perceived and received very differently.

An artist's process can also be seen as a kind of ambient poetics, as Timothy Morton noted, or of "making the imperceptible perceptible while retaining the form of imperceptibility."[14] Morton describes how the "Save the Whale" campaign used recordings of whale songs to draw attention to underwater environments inaccessible to humans, appealing to our aesthetic senses to relate to marine environments and their inhabitants without requiring us to understand the nature or meaning of those songs. The human ear listened, and the response—to new conservation strategies and marine protection policies—followed.

An artistic approach does not necessarily extend our senses to allow us to look deeper or further into something, as optical or acoustic instruments might do. Rather, it enables one to gain multisensory experience of things in relation. The color and smell of whale oil lamps' light or soot can be perceived differently today than they were in eighteenth-century London, because sensing comprises not only what is individually reconducted every time we sense something, but also mechanisms of distinction that are socially constructed over time. Londoners today and two hundred years ago have different modes of comparison, and different preconstituted potential objects of perception. Artistic experiments can, for example, present soot that is identical—yet not identical—to the soot in the air of eighteenth-century London, allowing us to both engage with the materiality of those conditions and simultaneously confront the fact that their full reconstruction remains impossible, always missing the eighteenth-century social-cultural bits that the act of perception activates. Similarly, there's a gap in perceptive capabilities between us and any other species, yet building experiments and experiences that aim to make tactile some of the impacts of the Anthropocene such as the increasing acidification and pollution of the seas bring us closer to and underscore our kinship with creatures that share some of those sensorial abilities with us.

14 Timothy Morton, *Ecology without Nature: Rethinking Environmental Aesthetics*. Cambridge, MA: Harvard University Press, 2007, p. 96.

Illumination

With the primacy of vision and of economic productivity, it is perhaps little wonder that the Anthropocene is ever more brightly lit.[15] Artificial light is one of the ways through which humans decouple themselves from nature, shunning the constraints of the Earth's rotation that plunges them into darkness every twelve hours, and enabling round-the-clock markets and infrastructures for continuous work and consumption. The journalist, librarian, and Indigenous rights activist Charles Lummis described in 1894 the Pueblo Isleta tale of why the Moon only has one eye.[16] Once upon a time, the Sun and Moon both lit the Earth with their bright eyes so that plants and trees could constantly grow, and humans and animals could have more time to work and play. They noticed that their children became weary and so the Moon sacrificed one of her eyes and allowed the other to slowly open and close, marking the phases of the lunar cycle and reminding humans of the gift of darkness, earthly rhythms, and sleep needed by all creatures. Today, the Moon's eyes have been opened again. In *24/7: Late Capitalism and the Ends of Sleep*, Jonathan Crary describes a 1990s proposal by a joint Russian–European space consortium to install reflectors in orbit which would illuminate the dark side of the planet. This proposal perhaps best exemplifies where the propensity for perpetual illumination might lead, how the drive for productivity and continuous circulation ends in "an instrumentalized and unending condition of visibility."[17] As human vision is increasingly more privileged both as a human sense and as an instrument of world-making projections, as it produces more complex visibilities and ever further-reaching captures of nature, as it traces and foregrounds carbon flows as generative and reparative, other senses, and other connections and material relations become more and more obscured.

15 Alejandro Sánchez de Miguel, Jonathan Bennie, Emma Rosenfeld, et al., "First Estimation of Global Trends in Nocturnal Power Emissions Reveals Acceleration of Light Pollution," *Remote Sensing*, vol. 13, no. 16 (2021), https://doi.org/10.3390/rs13163311

16 Summarized by Charles F. Lummis, *Pueblo Indian Folk-Stories*. London: Forgotten Books, 2008. Originally published by Charles F. Lummis in 1894.

17 Jonathan Crary, *24/7: Late Capitalism and the Ends of Sleep*. London: Verso Books, 2013.

With new forms of visibility come new forms of exposure and vulnerability. Artificial light alone is recognized by biologists as an anthropogenic pollutant, harmful to wildlife and humans (including potentially having contributed to the causes of the coronavirus (Covid-19)
pandemic[18] by disrupting the physiology and behavior of bats—the
nocturnal animals from which the virus "spilled out" to other species).
Despite these harms and dangers the use of artificial light is unregu-
lated, and its expansion or contraction is an incidental factor related
to managing the costs and efficiency of energy use. Light and energy
remain correlated, each driving the development of the other as they
have since the first human use of fire. And beyond disrupting dark environments, petroleum-based illumination plays a far larger role in
the degradation of habitats and environments as a catalyst and en-
abler of extractive and polluting modes of industrial and technological development, including contributing to computational technologies and media-disclosed visibilities that shape human understanding
of material relationships within environments.

       When twentieth-century systems ecology rendered natural en-
vironments as flows of matter of energy, and organisms as thorough-
ly embedded in networks of relations, it spotlighted those relations
which were productive to the ecosystem. Carbon was rendered as a
kind of energy currency which underpinned and unified material and
social systems. Today when we speak of "carbon," referring to both
the gaseous emissions in the air and the carbon sequestered as a climate remediation measure, we are guided by these legacies of understanding carbon as that which is conserved when carbon molecules
circulate through earthly spheres, taking on different embodiments
and forms. This way of thinking is structured by extractive political
economies that proclaim a sameness or commensurability underneath myriad carbon forms and processes. The idea of carbon as energy currency enables a new kind of commodification of nature
through carbon markets and an "accumulation through dispossess-
sion," where almost anything, even the vulnerable bodies of cetacean

---

18 Zeeshan Ahmad Khan, Thangal Yumnamcha, Gopinath Mondal, et al.,
     "Artificial Light at Night (ALAN): A Potential Anthropogenic Component
     for the COVID-19 and HCoVs Outbreak," *Frontiers in Endocrinology*,
     vol. 11 (September 2020), https://doi.org/10.3389/fendo.2020.00622

beings, could be calculated as a carbon sink or a carbon source to be exchanged. The carbon market itself becomes a system of control-through-calculation that casts the materiality of the biogeochemical world as pliant to human desires.[19]

Towards an Aesthetics of Relationality for the Anthropocene

Carbon has been used to render sameness as economic fungibility by reducing entities to one molecular dimension. But this shared chemical backbone could also be used to render connections across difference, connections that link forms of life and nonlife in what Angeliki Balayannis and Emma Garnett have described as "chemical kinship."[20] Entities emerge as materially bound to chemical relations. At the same time, they are formed by the molecular and relational diversity and complexity of chemical and ecological processes, based on the spectrum of bonds that the carbon atom makes with other elements and itself, each arrangement enabling a fantastic diversity of forms of life and nonlife and their emergent relationships. This is perhaps nowhere as poignant as the carbon reservoirs that are formed when a dying whale falls to the ocean floor. Reflecting on the complex relationships between the living and nonliving, if such categories can even exist in the Anthropocene, Julieta Aranda and Eben Kirksey ponder the "double death" recognizable in whale falls as "life becoming nonlife on a planetary scale."[21]

19 See, David Harvey, *The Limits to Capital.* London: Verso, 1982; see also Steffen Böhm, Maria Ceci Misoczky, and Sandra Moog, "Greening Capitalism? A Marxist Critique of Carbon Markets," *Organization Studies,* vol. 33, no. 11 (2012), pp. 1617–38.

20 Angeliki Balayannis and Emma Garnett coined the term "chemical kinship" in their 2020 article "Chemical Kinship: Interdisciplinary Experiments with Pollution," *Catalyst: Feminism, Theory, Technoscience,* vol. 6, no. 1 (2020): Special Section on "Chemical Entanglements: Gender and Exposure." We draw on their ideas to imagine chemical kinships with carbon.

21 Julieta Aranda and Eben Kirksey, "Toward a Glossary of the Oceanic Undead: A(mphibious) through F(utures)," *e-flux,* no. 112 (October 2020), https://www.e-flux.com/journal/112/354965/toward-a-glossary-of-the-oceanic-undead-a-mphibious-through-f-utures/

The initial rendering of the natural and social environments as networks of relations in an ecosystem was followed by a proliferation of ecological reframings: ecologies of media, perception, power, or information, to name a few. The Anthropocene has come to be marked by what we might call an "aesthetics of relationality." While relationality is a lens favored across the ideological spectrum, its roots in the discipline of ecosystems ecology carry into contemporary posthuman thinking the paradigms of technological and algorithmic forms of control, as Erich Hörl notes. Far from being just a metaphor, ecology as an analytical paradigm in any sphere "reflects the thoroughgoing imbrication of natural and technological elements in the constitution of the contemporary environments we inhabit."[22] Judith Butler also warns that the ecological metaphor and lens do not guarantee that the kind of relations they make visible are not causing harm. She writes that "relationality is not by itself a good thing, a sign of connectedness, an ethical norm to be posited over and over again against destruction; rather, relationality is a vast and ambivalent field in which the question of ethical obligation has to be worked out in a light of a persistent and constitutive destructive potential."[23] What is "worked out" of relational fields, and through what means, is critical for the politics of nature and coexistence. The insights gained in an aesthetic register allow us to foreground not only the more diverse and unseen kinds of relations but also what happens across them: the vulnerability, responsibility, separations, fears, or collectivisms required and produced by these profound interdependencies.

Our relationship to the environment is always mediated, based on concepts and representations, in the production of which technologies play just as much a part as sociocultural processes. Artistic approaches that open this pretext of our understanding by asking silly or fantastical questions can expand the context of meaning to include

---

22 Erich Hörl, "Introduction to General Ecology," in James Burton and Erich Hörl (eds), *General Ecology: The New Ecological Paradigm*. London: Bloomsbury, 2017, pp. 1-73.

23 Judith Butler, *The Force of Nonviolence. An Ethico-Political Bind: The Ethical in the Political*. London: Verso, 2021, n. p., https://iberian-connections.yale.edu/wp-content/uploads/2020/09/The-Force-of-Nonviolence-An-Ethico-Political-Bind-by-Judith-Butler.pdf

new sensual experiences, new ways of reestablishing a relationship to that which surrounds us. This approach may introduce a shift away from clear definitions and representations towards the intensities, differences, and indeterminacy that are part of them. Living in a time in which our understanding of vegetal and animal life, planetary processes, and the role of human influence on a planetary scale are constantly being reshaped, the necessity of conceptual knowledge in aesthetic theory that goes beyond the boundaries of the sensually perceivable seems necessary. Facing the limits of human comprehension in these matters allows us to attribute new forms of agency and sentience to our environments and the creatures we share these environments with.

Perhaps the best use of the lens of aesthetics is to turn it on ourselves to investigate our responses to different registers of whale encounters. To our twenty-first century ears, the twentieth-century descriptions of hooking and harpooning whales, full of gruesome details of how long whales took to die in pools of their own blood, sound sickening. Erin Hortle suggests that this sickening feeling might be lurking behind our new love for this species, among other reawakened kinships with other creatures.[24] "Is this love based on an awareness of the guilt of the collective human?" Hortle asks. Does it respond to the feeling of shame brought about by a visceral experience of "getting it"—in the words of Thom Van Dooren[25]—that is, of "getting" the enormity of human impact across time that brought these species to near extinction? While attempting to define or find the markers of the Anthropocene, aesthetic approaches might be necessary if this new designation of our geological era seeks to orient us to new, less-destructive approaches for coexisting on the planet.

In the natural-cultural history of the whale and its material labor for human well-being and economic productivity, we can see a mechanism that is deeply inscribed into the relationship of the human being to its environment in the Anthropocene: an obscuring of a vast field of material relationships in favor of the visualization and amplification

---

24 Erin Hortle, "Historicising Ambergris in the Anthropocene," *Australian Humanities Review*, 63 (November 2018), pp. 48–64.
25 Thom Van Dooren, *Flight ways: Life and Loss at the Edge of Extinction*. New York: Columbia University Press, 2014.

of those beneficial to the needs of humans. The question posed by the Carbon Aesthetics group is what role aesthetics can play in order to make other senses of material interrelations traceable, and to point to other and new relationships—not just relationships that have a directly recognizable connection to anthropogenic value systems, but extending to ones that reach far beyond them, showing that the human being is embedded in processes and infrastructures that cannot be understood, operationalized, or perceived in their totality.

This contribution emerges from the conversations and visual research of a group of artists and humanities scholars who explore carbon and its relations under an umbrella project of the Carbon Catalogue. The Carbon Catalogue is an artistic research project led by Karolina Sobecka, which explores the contemporary preoccupation with carbon as a by-product of anthropocenic fossil fuel combustion. The Carbon Catalogue documents new designations, technologies, and imaginaries of carbon, while also putting them in play, reordering and restorying what carbon is becoming.

Carbon Aesthetics, a working group of the Carbon Catalogue, is organized by Desiree Foerster and brings together artists who turn to the material and the elemental, as well as to the symbolisms and common representations of the interrelations between biochemical and technological processes inherent to the carbon cycle. Artists in the group are Andrés Burbano, Myriel Milicevic, Alex Toland, and Clemens Winkler. The group members foreground emphasis on the relations that constitute objects, which allow the movement of materials, and, thus, form the infrastructure for our meaningful interactions with the world.

Coral Times

Kat Austen: Coral has a wide history of entanglements with humans. There is a wealth of historical references to coral from all kinds of perspectives, from all kinds of people.

Nigel Clark: In William Shakespeare's *The Tempest*, there is mention of corals using human parts as construction materials. In 1842, Charles Darwin published his study of atolls, linking coral reefs to volcanoes. In 1845, Karl Marx writes this famous line, saying that we have already transformed the entire surface of the Earth "except perhaps on a few Australian coral-islands of recent origin." Throughout history, there is this thinking with coral. All sorts of different things of value, the mutualism, the working with other species, the idea that Marx mentioned about coral being inherently collectivist. You can project all sorts of things onto corals.

But there is still a big question for me about what we might learn from corals. And the fact that they have endured huge climatic change, huge sea-level changes, the moving of continents. They not only survived these adversities; they thrived. We can learn from the corals, not just for the sake of reef restoration, but for the way we think about and do our own constructions, build our own cities.

It is kind of bizarre that we are trying to restore corals, but we are simultaneously still building huge settlements, our own biogenic rocks by the coast, knowing full well that the sea level is going to rise and literally take those cities and turn them into reefs. What kind of interaction might we expect in the future if the corals do survive? How are they going to interact with our cities as our cities go underwater? What are they going to make of the biogenic structures that we leave for them in their ocean?

Kristine DeLong: This makes me think of diving around shipwrecks. In places like Bermuda, the corals will start to grow on sunken ships and encrust the old metal surfaces and things that get left behind. Eventually, the corals form a new reef.

Jens Zinke: There are several islands in the Pacific that are currently under threat of drowning. We can probably watch this happening over the coming decades, which is terrible for the people who live on these islands. They will be forced to leave their home at some stage, and they're trying to find a new home. This comes with all kinds of repercussions; the corals will probably have the opportunity to resettle on some of the things that get drowned. But it could also happen that all the traces that we leave behind, our pollution and all the things that we produce, get flushed into the waters, and the sediment gets eroded from the islands and pollutes the lagoons.

It is also possible that the corals will thrive and regrow, but this might take longer. It happened already after the last Ice Age, during which several meters of the coral reefs were sticking out of the water. When the oceans flooded again, it took some time for the corals to kind of find themselves. They were hiding somewhere further down the slope, where there was water. Then, slowly but surely, they resettled. But it took several hundreds of years until they actually reestablished themselves on the old reefs where they had grown thousands of years before. It is not a quick process; it actually takes quite a lot of time.

NC: The fascinating thing for me about geology is the way that things get sedimented. How materials sink down and then resurface someplace else. I was reading something about atavistic corals, that there may be some corals around that are better adapted to the past conditions of the Earth, and that they're hopefully still around.

It's fascinating, because these corals exist in multiple temporalities. Take a mantle plume, a mass of material in the Earth's mantle, some of which rises to the surface as the ocean bed slowly moves across it. So from this, you get a volcano, another few million years later, and another volcano after that in a line, each of them attracting corals. In a sense, the corals are inhabiting very, very different temporalities at the same time. Different corals might almost be inhabiting different times, or at least, have their own very different tempos.

KDL: Millions of years ago, we used to have a warm ocean current that went all the way around the Earth's equator. It went from the Indian Ocean, Pacific Ocean, into the Mediterranean—or what is today the Mediterranean, it was a different sea at the time—into the Atlantic, the Caribbean, then through Panama, which was open back then. During this time, coral diversity was pretty much the same everywhere, because the young coral polyps traveled along this warm ocean current. From when the connection between the Pacific and the Atlantic Ocean in Panama closed, we see this drop off in biodiversity in the Atlantic Ocean. We have been going through this kind of prolonged extinction event you see, so the reefs we see today in the Caribbean are very different from the reefs in the Pacific.

A few years back, I was in Indonesia, and I hadn't been to an Indo-Pacific reef in some time. I put my head underwater and was like: Oh my gosh, this is so gorgeous. There are so many corals. Diving in the Atlantic, you just do not have that same diversity. There are thousands of different coral species in the Pacific, while we are down to about thirty-six in the Atlantic. And the ones that remain are the survivors—these are the ones that have adapted to all these changing conditions and the changing salinity in the Atlantic Ocean. And now with anthropogenic climate change, we are kind of rushing these Atlantic species towards their extinction.

The coral that I'm working on for the Anthropocene GSSP [Global Boundary Stratotype Section and Point] project, *Siderastrea siderea*, is a survivor. You could put sediment on it, and it will survive. In the Dry Tortugas off the coast of Florida, there was huge coral die-off in the 1880s. Therefore, all the coral colonies there are about the same age, except for this one massive *Siderastrea siderea*, a coral that survived that event over one hundred years ago. So, the species I'm working with now, these are the survivors, these are the strong ones.

The above exchange is an edited excerpt from "Conversations Beyond the Human," a discussion held on May 20, 2022 at HKW in Berlin during the event *Unearthing the Present*.

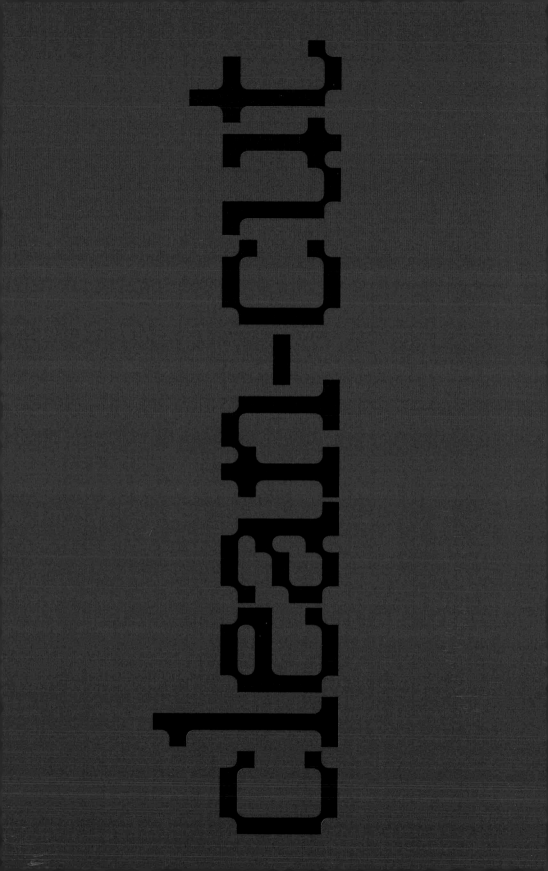

clean-cut

Anthropogenic Fire as the Hinge between Earth System and Strata

Acceleration, Deceleration

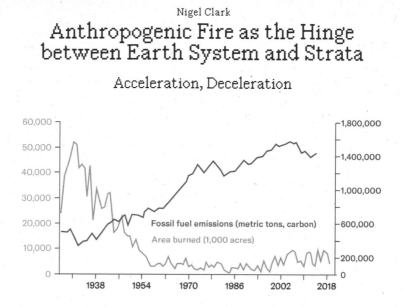

Midway through his career-crowning text on the history of human fire use, Stephen Pyne presents a graph charting the course of two different kinds of combustion in the United States from the early twentieth century to the present.[1] The curve of fossil fuel emissions, unsurprisingly, climbs steadily upward. Though bumpier, the other curve—the total area annually burned by wildfire—descends as dramatically as its counterpart rises.

Relevant well beyond the US, the graph neatly captures the point Pyne has been making for over three decades: in the process of ascending to global climate-altering levels, anthropogenic combustion of fossil hydrocarbons has displaced another kind of fire. And the quenching of that other fire—the burning of living or recently living biomass—he argues, is just as significant as the unleashing of combustible matter from its lithic reservoirs. The monstrous fires that have erupted in recent years in Australia, California, the Mediterranean, and many other pyrophytic regions of the planet, Pyne insists, are evidence that substituting the closed fire of fossil-fueled heat engines for the open fire of landscape burning is utterly unsustainable.

1 Stephen Pyne, *The Pyrocene: How We Created an Age of Fire and What Happens Next*. Oakland, CA: University of California Press, 2021, p. 103, fig. 3.

Wildfire is photogenic, providing the visual and visceral appeal missing in so many other depictions of changing climate or shifting Earth systems. But this scintillating media presence conceals the fact that, despite escalating megafires, there is far less fire in the planetary landscape than there was half a millennium ago.[2] Overall, then, what we are witnessing is a deficit not an excess of burning biomass, or as Pyne puts it, the Earth now has "too much bad fire, too little good fire."[3]

Why might this be important for the identification of markers for the Anthropocene? Formalization of the hypothesis matters. But we shouldn't forget the claim by its leading exponents that the Anthropocene "has the capacity to become the most politicized unit, by far, of the Geological Time Scales—and therefore to take formal geological classification into uncharted waters."[4] No less than the choice of a starting date, the selection of markers has potentially profound implications for how we understand, distribute, and reimagine human agency.

We should also recall that the question of a possible departure from Holocene conditions arose out of the relatively new field of Earth system science—prior to the turn to the more established discipline of geology for confirmation. The subsequent collaboration between the study of "hard rock" geology and the more mobile envelope of the outer Earth system may itself mark a significant juncture in the scientific understanding of how the Earth operates. As Zalasiewicz et al. explain, "Geologists [...] benefit from this mutual exchange [...] as it enables better process models of the stratigraphical data," while benefits to Earth system science accrue from "the recognition of geological signals as additional data and proxies [...] especially for testing models and forecasting future scenarios."[5] Consequently, the Anthropocene

2   Stefan H. Doerr and Cristina Santín, "Global Trends in Wildfire and its Impacts: Perceptions versus Realities in a Changing World," *Philosophical Transactions of the Royal Society B,* vol. 371 (2016), https://doi.org/10.1098/rstb.2015.0345, accessed October 26, 2022. All online references in this essay were last accessed in October 2022.
3   Pyne, *The Pyrocene*, p. 5.
4   Jan Zalasiewicz, Mark Williams, Will Steffen, and Paul Crutzen, "The New World of the Anthropocene," *Environmental Science and Technology*, vol. 44, no. 7 (2010), pp. 2228–31, here p. 2231.
5   Jan Zalasiewicz, Will Steffen, Colin Waters, et al., "Petrifying Earth Process: The Stratigraphic Imprint of Key Earth System Parameters in the Anthropocene," *Theory, Culture & Society*, vol. 34, no. 2/3 (2017), pp. 83–104, here p. 97.

hypothesis may already be shifting the platform on which it seeks to ground itself.

Fire is an especially potent intermediary between the Earth system and the lithic strata, I want to suggest, and the human capture of fire is key to our species' acquisition of geological agency.[6] This brings us back to Pyne's intersecting downward inflection of landscape burning and upward arc of fossil fuel combustion—curves that tellingly part company, at least in the US case, around 1950. By comparison with the "Great Acceleration" of fossil-fuel combustion, the "great deceleration" of landscape fire may be too discontinuous and difficult to disaggregate from other signals to be an independent contender for marking the Anthropocene. On the other hand, too hastily severing the ascent of fossil hydrocarbon combustion from the descent of landscape burning may well preemptively tease apart what Anthropocene science has so promisingly woven together.

This becomes even more pertinent when we consider the possible political repercussions of electing an Anthropocene marker. To date, the chosen markers foreground predominantly Western technological developments, the rhetoric of "acceleration" itself mirroring the industrial capitalist axiomatic of continuous linear accumulation. In the process, other modes of inhabiting the Earth, in particular the use of techniques and practices that impact longitudinally rather than synchronously, risk being obscured.[7] This is particularly problematic in the case of nuclear testing, much of which took place in the unceded customary land, sea, and air of colonized peoples. Related risks attend highlighting the signatures of combusting fossil hydrocarbons, if treated in isolation. But if we at least attempt to address the decline of landscape burning in tandem with rising fossil fuel combustion, a deeper and much more shared history of cultural burning comes into relief.

In the following, extending Pyne's notion of a "pyric transition" from open-field landscape burning to the chambered combustion of

6   Nigel Clark and Lauren Rickards, "An Anthropocene Species of Trouble? Negative Synergies between Earth System Change and Geological Destratification," *Anthropocene Review* (June 28, 2022), https://journals. sagepub.com/doi/epub/10.1177/20530196221107397

7   See Matt Edgeworth, Erie C. Ellis, Philip Gibbard, Cath Neal, and Michael Ellis, "The chronostratigraphic method is unsuitable for determining the start of the Anthropocene," *Progress in Physical Geography: Earth and Environment*, vol. 43, no. 3 (2019), pp. 334–44.

fossil biomass, I suggest how an expanded focus on fire not only weaves longitudinal and globally synchronous forms of human geological agency into one narrative, but also strengthens the conceptual convergence of the study of Earth systems and the lithic strata.

Fire as Marker

Fire is the vernacular term for a rapid, positive feedback reaction that converts chemical energy into thermal energy. While some other astronomical bodies in our solar system have the ingredients of fire, Earth is the only planet on which the necessary components of fuel, ignition source, and an oxidizing agent are fully integrated.[8] Here, photosynthesizing life-forms turn sunlight into energy-rich carbon compounds, while fire reverses the equation by decomposing carbon-rich organic matter into thermal energy. The simple presence of life, however, is not enough. It took a planet-wide oxidation event, the rise of multicellular organisms, and the colonization of land by plants to finally fuse fire's three ingredients, possibly beginning in the early Devonian.[9] It took another 400 million years or so for the "fire planet" to evolve a creature capable of handling fire.

For most of the million or more years that the extended human family has been manipulating fire, the impact has been localized, intermittent, and patchy. The proposal that anthropogenic fire-enabled deforestation deep in the Holocene helped defer the return of an ice age is contentious but has yet to be ruled out.[10] A stronger early contender for the onset of planet-wide anthropic impact, initially favored by Anthropocene progenitor Paul Crutzen, was the take-off of fossil-fueled industrialization—"the thermo-industrial revolution of nineteenth-century Western civilization."[11] As the demand for a pronounced

8   Stephen Pyne, *Fire: A Brief History*. Seattle, WA: University of Washington Press, 2001, p. 3.
9   Ibid.
10 Andrey Ganopolski, Ricarda Winkelmann, and Hans Joachim Schellnhuber, "Critical Insolation–CO2 Relation for Diagnosing Past and Future Glacial Inception," *Nature*, vol. 529, no. 19/20 (2016), pp. 200–07.
11 Will Steffen, Jacques Grinevald, Paul Crutzen, and John McNeill, "The Anthropocene: Conceptual and Historical Perspectives," *Philosophical Transactions of the Royal Society A*, vol. 369, no. 1938 (2011), pp. 842–67, here p. 847.

geosynchronous signal turned attention to the post-Second World War surge of globalization, it is the spheroidal carbonaceous particle (SCP) that has most clearly inherited and updated the "thermo-industrial" thematic.

A subset of fly ash—airborne particulate by-products of high-temperature fossil fuel combustion—SCPs are residues of the incomplete burning of pulverized coal or oil droplets.[12] While their microscopic size contributes to global atmospheric diffusion, SCPs also have the advantage of having no natural counterpart and thus being readily distinguishable in sedimentary samples. But in making a case for SCPs as an especially robust indicator of a mid-twentieth-century Anthropocene onset, Neil Rose goes beyond emphasizing their ubiquity and convenience, stressing their link to a fundamental driver of anthropogenic global change: the combustion of fossil fuels.[13]

In the bigger picture of Anthropocene science, this direct connection between the physical archive of the novel lithic strata and significant Earth system change seems to offer something lacking in the case of the proposed radionuclide marker—which for all the clarity of its signal has a much more ambiguous connection with human impact on Earth processes. However, if we step further back, the shared, integrative force of fire begins to show up in other proposed markers. In one way or another, anthropogenic fire underpins rising atmospheric and oceanic carbon dioxide concentrations, the broader continuum of black carbon deposits, and proliferating human-made minerals such as concrete, alloyed metals, glass, ceramics, and plastics.

So we should also consider the significance of fire in early iterations of the idea that human activity might transform Earth processes in their entirety. In a 1982 paper, Crutzen conjectured that a nuclear exchange would result in massive wildfires generating photochemical smog that could "change the heat and radiative balance and dynamics of the earth and atmosphere" with devastating impact on surviving

---

12 Neil L. Rose and Agnieszka Gałuszka, "Novel Materials as Particulates," in Jan Zalasiewicz, Colin N. Waters, Mark Williams, and Colin P. Summerhayes (eds), *The Anthropocene as a Geological Time Unit*. Cambridge: Cambridge University Press, 2019, pp. 51–58, here p. 51.

13 Neil L. Rose, "Spheroidal Carbonaceous Fly Ash Particles Provide a Globally Synchronous Stratigraphic Marker for the Anthropocene," *Environmental Science & Technology*, vol. 49 (2015), pp. 4155–62, here p. 4160.

humans.[14] More generally, fire came to play an integrative role in Crutzen's vision of a dynamic and changeable Earth system. It's also worth
recalling his early efforts to distinguish between forms of combustion
that added carbon to the atmosphere and those that returned carbon
to the soil—notably the biomass burning of shifting cultivators.[15] As
Crutzen and his co-author later concluded in a collection that integrated the fields of wildland fire science and atmospheric chemistry:
"the preservation and study of fire will assist humanity in its larger
stewardship of the Earth."[16]

Taking inspiration from both Pyne and Crutzen, I want to step
back still further from the question of identifying an end-of-Holocene
marker in order to dig deeper into the issue of how Anthropocene science can help us make sense of human planetary agency. Just as fire,
over the last 400 million years, has played a vital part in the interactions between the relatively mobile envelopes of the outer Earth system and the slower-moving fabric of the lithic strata, it is the capture
of fire by humans, I suggest, that has enabled us to emerge as a particularly active hinge between these two planetary domains.

Human Fire at the Strata-Earth
System Juncture

Terrestrial fire is predominantly a surface phenomenon. Many organisms take advantage of fire—to open seeds, promote new growth,
flush out prey—but only humans actively manipulate flame. More
than an event in *human* history, Pyne insists, "the capture of fire by
Homo marks a divide in the natural history of the Earth."[17] If skillful

14 Paul Crutzen and John Birks, "The Atmosphere After a Nuclear War:
Twilight at Noon," *Ambio*, vol. 11, no. 2/3 (1982), pp. 114–25, here p. 123.
15 Wolfgang Seiler and Paul Crutzen, "Estimates of Gross and Net Fluxes of
Carbon Between the Biosphere and the Atmosphere from Biomass
Burning," *Climatic Change*, vol. 2 (1980), pp. 207–47.
16 Johann Goldammer and Paul Crutzen, "Fire in the Environment:
Scientific Rationale and Summary of Results of the Dahlem Workshop,"
in Johann Goldammer and Paul Crutzen (eds), *Fire in the Environment:
The Ecological, Atmospheric, and Climatic Importance of Vegetation Fires*.
Chichester: Wiley, 1993, pp. 1–14, here p. 11.
17 Stephen Pyne, "Maintaining Focus: An Introduction to Anthropogenic
Fire," *Chemosphere*, vol. 29, no. 5 (1994), pp. 889–911, here p. 889.

landscape burning has dramatically increased the range and niche of humans, however, so too has domesticated fire been the key to the human traversal of the Earth "vertically."[18]

As diurnal, surface-dwelling creatures, we need flames to light the way underground. It may not be coincidental that our distant ancestors look to have acquired the ability to handle fire in an environment where they also negotiated dynamic and fractured rock formations. East Africa's Rift Valley—the largest, most long-lived fracture zone on the Earth's surface—is characterized by "complex tectonics and intense volcanism."[19] Rift Valley topography was conducive to frequent patchy burning, while its constant volcanic activity supplemented lightning's spark, and there has long been speculation that hominins first captured flame not from raging wildfire but from the more constant ebb of lava in their immediate environments.[20] There are also intriguing signs that, having migrated away from ancestral volcanic homelands, ancestral humans learned to bury stones beneath hearth fires—using heat to transform available sedimentary rock so it acquired some of the flaking and sharpening properties of volcanic rock.[21] If this is the case, then already 70,000 years ago, humans were using high heat to restructure inorganic matter—and in the process, reconfiguring their relationships with the subsurface.

This fire-mediated articulation between the Earth's surface and the rocky strata intensifies with the enclosure and intensification of flame. The earliest purpose-built fire containers—rudimentary kilns excavated at Dolní Věstonice—are estimated to be 26–30,000 years old.[22] When the final Pleistocene glaciation ceded to warmer, steadier climates and some nomadic peoples settled into more sedentary lifestyles, chambered fire burgeoned into a vital constituent of Neolithic

18 Nigel Clark, "Vertical Fire: For a Pyropolitics of the Subsurface," *Geoforum*, vol. 127 (December 2021), pp. 364–72.
19 Geoffrey King and Geoff Bailey, "Tectonics and Human Evolution," *Antiquity*, vol. 80, no. 308 (2006), pp. 265–86, here p. 277.
20 Clark, "Vertical Fire."
21 Kyle Brown, Curtis Marea, I. R. Herries, et al., "Fire as an Engineering Tool of Early Modern Humans," *Science*, vol. 325 (September 2009), pp. 859–62.
22 Pamela B. Vandiver, Olga Soffer, Bohuslav Klima, and Jiři Svoboda, "The Origins of Ceramic Technology at Dolní Věstonice, Czechoslovakia," *Science*, vol. 246 (November 1989), pp. 1002–08.

life. Ovens rendered grains digestible, and out of kilns came earthen-ware vessels, bricks, tiles, and later, metals and glass.[23]

The search for metallic ores drew us further into the depths of the Earth, and mining made new demands of fire. "Fire-setting"—exposure to high heat followed by quenching—was early miners' chief means of cracking rock. "Prospectors burned over hillsides to expose rock," recounts Pyne; "Miners relied on fire to tunnel, to smelt, to forge."[24] As mining fed ores into the furnace, tools forged by metalworkers expediated extraction, and as demand for ores escalated, the drive and ability to extract these minerals correspondingly advanced. Again, we can see the enclosed fire of the artisanal furnace as a novel hinging together of mineral-bearing strata and Earth system fluxes.[25] In the ancient Middle East, as archaeologists document, there was a dynamic, self-reinforcing trade relationship between highland metallurgy and the intensive grain cultivation of the alluvial lowlands[26]—or what we might view as a new articulation between sedimentary and metalliferous zones.

Although Pyne himself refers to the longer history of chambered fire, there is a sense in which 20,000-plus years of pyrotechnology complicates his more singular notion of a pyric transition between fossil-fueled heat engines and landscape burning. A further complication comes with the invention of another kind of fire: the positive-feedback biochemical reaction sped up to a split-second.

Over the course of extensive experimentation, researchers in ninth-century China pioneered a form of combustion in which the sudden release of pure oxygen accelerates the conversion of available fuel into hot gas in a few thousandths of a second. While the geological impact of escalating firepower has been noted, less attention has been given to understanding weaponized explosions as applications of a novel kind of fire.[27] Indeed, we might see near-instantaneous

23 Theodore Wertime, "Pyrotechnology: Man's First Industrial Uses of Fire," *American Scientist*, vol. 61, no. 6 (1973), pp. 670–82.

24 Pyne, *Fire: A Brief History*, p. 131.

25 Nigel Clark, "Fiery Arts: Pyrotechnology and the Political Aesthetics of the Anthropocene," *GeoHumanities*, vol. 1, no. 2 (2015), pp. 266–84.

26 Leslie Aitchison, *A History of Metals, Volume 1*. London: MacDonald & Evans, 1960, pp. 18–24.

27 For the geological impact, see Mat Zalasiewicz and Jan Zalasiewicz, "Battle-Scarred Earth: How War Reshapes the Planet," *New Scientist* (March 25, 2015), https://www.newscientist.com/article/mg22530140-600-battle-scarred-earth-how-war-reshapes-the-planet/

combustion as the first entirely new form of fire on Earth for over 400 million years: a great acceleration of combustion that both anticipates and enables key aspects of the better-known post-Second World War "Great Acceleration."

Explosive gunpowder and its successors also have significant nonmilitary impacts on the mixing or turbation of rock fabrics. By the mid-nineteenth century, commercial applications of gunpowder for mining and civil engineering had overtaken military uses.

As well as these direct geological impacts, ultra-high-speed combustion has indirect but momentous repercussions through the historical linkage between explosive weapons and the heat engines that powered industrialization. As Lewis Mumford observed in the 1930s, "the gun was the starting point of a new type of power machine: it was, mechanically speaking, a one-cylinder internal combustion engine."[28] Joseph Needham fills out this storyline—tracking a history of schemes and experiments to put gunpowder to useful work that go back to the sixteenth century. Scientist-inventor Christiaan Huygens' project with the French Academy of Sciences in the 1670s is pivotal. As Huygens wrote: "The force of cannon powder has served hitherto only for very violent effects [...] people have long hoped that one could moderate this great speed and impetuosity to apply it to other uses."[29]

Initially working under Huygens on the *moteur à explosion*, it was Denis Papin who recognized that steam power offered a "less violent" route to creating the vacuum that could drive a piston. Papin set research and development on a path towards external combustion—using fire-heated boilers as a motive force. Though not powered with gunpowder, the internally combusting *moteur à explosion* would be revived some two centuries later as the driving force of the automobile.[30] The fossil-fueled automobile, in turn, would add its immense heft to the shifting relationship between the lithic strata and the Earth system—adding weight to the idea that a chain of pyric transitions lies

28 Lewis Mumford, *Technics and Civilization*. London: George Routledge & Sons, 1934; Chicago, IL: University of Chicago Press, 2010, p. 88.
29 Quoted from Joseph Needham, *Science and Civilisation in China: Volume 5*. Cambridge: Cambridge University Press, 1986, p. 557.
30 Nigel Clark, "Infernal Machinery: Thermopolitics of the Explosion," *Culture Machine, vol. 17: Thermal Objects* (2019), http://culturemachine. net/vol-17-thermal-objects/infernal-machinery/

behind successive transformations in the human capacity to hinge together the Earth system and the lithic strata.

Combustive Justice and the Anthropocene

Following the "wide initial approach" to the Anthropocene, the demands of formalization call for the selection of a "primary" signal:
a single reference point deputizing for the breadth of anthropogenic
impacts on Earth processes and structures.[31] A careful, judicious
framing is required if this obligation towards a certain reductiveness
is not to be politically counterproductive. No less, care must be taken
so that Anthropocene science's most radical maneuver—its fusion of
"hard rock," deep time geology, with Earth system science—is not to
be pushed into the background. In this final section, I make the case
that a more explicit concern with the "integrative" thematic of human
fire use could help us achieve both these aims at the same time.

There is great potential for the convergence of "stratigraphic"
and Earth system thinking to open new perspectives on the way our
species and its extended hominin family has gradually accrued its
planet-altering agency. Such an approach, as I've been illustrating,
helps us to see how the diverse setting-to-work of fire has played a key
role in human intervention in the flows and cycles of the Earth system,
in their traversal of lithic strata, and in their hinging together of these
two planetary domains. If fire, as Pyne insists, *integrates* different environmental processes, so too might we say that it *articulates* between
the major structural divisions of the Earth.[32]

Indeed, we might push this idea further. As Earth system scientists remind us, "the planet Earth is really comprised of two systems—
the surface Earth system that supports life, and the great bulk of the
inner Earth underneath."[33] Through the containment and intensification of fire, the genus that emerged in and around the volcanically

---

31 On the "wide approach," see Jan Zalasiewicz, Colin Waters, Colin
    P. Summerhayes, et al., "The Working Group on the Anthropocene:
    Summary of Evidence and Interim Recommendations," *Anthropocene*,
    vol. 19 (September 2017), pp. 55–60, here p. 56.
32 Pyne, *The Pyrocene*, Prologue.
33 Tim Lenton, *Earth System Science: A Very Short Introduction*. Oxford:
    Oxford University Press, 2016, p. 17.

active Rift Valley has learned to reproduce some of the forces of the inner Earth. Already by 6,000 BP, high-heat artisans were stoking their furnaces to 12–1300 degrees Celsius—a temperature that approximates the maximum heat of lava.[34] By using their kilns to melt and recrystallize rock, to metamorphosize minerals, to decompose and concentrate metallic ores, they effectively enfolded some of the power of the subcrustal Earth into the everyday spaces of their villages and towns.[35]

Later, with the weaponizing of gunpowder into explosive devices, Chinese military engineers set in play a mobilization of matter so rapid that it overtook even the 2–300 meters per second velocity of rocks ejected during volcanic explosions.[36] If this new fire dramatically accelerated the exchange between the Earth system and strata, it also played a preparatory role for the thermonuclear explosion—which in a certain sense domesticates the nuclear fusion processes that power stars such as our own Sun. In this regard, the fiery explosion can be seen as a step toward another kind of hinging together between systems—this time terrestrial and cosmic: a point that might be extended towards rocket propulsion and the ability to leave the Earth's orbit.

Whether by way of fire or other elemental processes, I suggest that new framings of the Earth system-strata interface help us to understand how humans acquired planetary agency. But this is also a matter of justice. It is an issue of acknowledging the multiple ways that different human collectives—throughout history and across the globe—have engaged with a dynamic, richly-resourced planet; and it is a question of confronting the suppression and marginalization these traditions have so often faced.[37] Just as researchers talk about a "black carbon continuum" in reference to the many ways that human

34 J. E. Rehder, *The Mastery and Uses of Fire in Antiquity*. Montreal: McGill-Queen's University Press, 2000, p. 54.
35 Clark, "Vertical Fire."
36 Gary Settles, "High-speed Imaging of Shock Waves, Explosions and Gunshots," *American Scientist*, vol. 94, no. 1 (2006), https://www.americanscientist.org/article/high-speed-imaging-of-shock-waves-explosions-and-gunshots
37 Nigel Clark and Bronislaw Szerszynski, *Planetary Social Thought: The Anthropocene Challenge to the Social Sciences*. Cambridge: Polity Press, 2021.

agents generate residues from combustion, so too do we need to think of a broader continuum in which all extant human populations and many of our hominin ancestors played a significant role in learning how to negotiate planetary variability using fire and other forces.[38]

Whereas focusing on a radionuclide signature may be clear and "unambiguous" in important regards, it also risks masking the agency of Aboriginal Australians, Pacific Islanders, Kazakhs, and others on whose customary lands weapons testing so often took place. In the case of Indigenous Australians, this could occlude tens of thousands of years of shaping an entire continent through skilled application of fire to living ecosystems—in this way risking a return to the racist imaginary of "primitive" people on the receiving end of unfathomable Western technological supremacy.[39]

Conversely, addressing a continuum of pyrogenic impacts that treats the signature of marginalized and ascendant practices as two sides of a definitive, shared anthropic attribute might signpost a greater willingness "to take formal geological classification into uncharted waters." But pursuing combustive justice is not simply a question of ceding objectivity to political imperatives. It is also about directing scientific attention towards processes that have emerged and developed over thousands, tens, or even hundreds of thousands of years: a matter of digging beneath comparatively shallow stratigraphic signatures to unearth their more profound conditions of possibility. We may well learn some valuable lessons by registering the traces of nuclear test ban treaties, but only by exploring the deep, complex, and tangled human history of intervening in elemental processes will we gain an appreciation—or reappreciation—of alternative possibilities for joining forces with the Earth.

38 On the "black carbon continuum," see Rose and Gałuszka, "Novel Materials as Particulates," p. 52.
39 See Marcia Langton, *Burning Questions: Emerging Environmental Issues for Indigenous Peoples in Northern Australia*. Darwin: Northern Territory University, 1998.

Violet.

How Political is the Stratigraphic Anthropocene?

Simon Turner: You cannot separate science from politics. I think we all pretend that we can, but we understand that institutions are not purely objective. There are political dimensions to science. See for instance the anthropogenic drivers that we are discussing: there is a social, economic, and historical reason why the Chinese signal of carbonaceous particles produced by high temperature fossil fuel consumption occurs later than the one you see in North America—it's down to the historical uptake and expansion of power station technology; clearly a relation to very different political systems operating in the twentieth century. These sorts of things are understood. Some of the criticisms about the AWG revolve around saying that we, the geologists, should not be involved with this, because we should not be trying to define something that is politically such a hot potato and outside of our discipline.

Imagine, as a political scientist, that we decide next week that the Anthropocene is real. We draw the line, we push it through, and the ICS [International Commission on Stratigraphy] agrees. "Cool. Thanks for all your effort, it is now the Anthropocene." What impact does that actually have on political thought? Well, I love the idea of being in the UK Parliament's House of Commons: "Thank you, Prime Minister. Thank you. Thank you. We've officially entered the Anthropocene now." But I believe that most people in politics would not even blink. What do you think?

Victor Galaz: I would say we should look back on the past ten years and how the Anthropocene as a concept has been treated since it was introduced. And from a political perspective, you can see that there are very different narratives being built around the Anthropocene. One of these narratives could be that we've moved into this new epoch where humans are driving climate change: "Look at this big impact that we're having and the massive risks. We need to put the brakes on climate change." Another narrative could be that humans have always been an ingenious species; we've always transformed ecosystems and the biosphere to benefit ourselves, and this is just further evidence of how remarkable we are. All we have to do is to become better at governing this new epoch.

These are quite different stories around the Anthropocene. I don't think that this is going to change just because you have a formal decision. I'd argue that the level of understanding of the Anthropocene, among the public and among environmental and climate policy-makers, is extremely shallow. Whenever you talk about the Anthropocene in public, even with people who are involved in climate policy, they will just talk about climate. They will say: "Anthropocene equals climate change; we need green energy." For me, that's just not the right way to look at it.

ST: I believe that maybe the scale of the Anthropocene is not yet properly understood. In order to address the challenges and disruptions of the Anthropocene, we need societal changes comparable to and greater than those implemented during the recent pandemic, and I would hope that a decision on an epochal scale might help to make that clear. But then again, the search for a GSSP [Global Boundary Stratotype Section and Point] in itself is not a political project.

VG: I don't think that's how it is viewed from the outside. I think there is a difference between the most important political implications of the work conducted by the AWG, and the conception of this being a political process in itself. Those are two very different things. You can acknowledge that you are part of a political context, but still maintain that this is a scientific process.

The above conversation is an edited excerpt from "Exchange on Collaboration and Complexity," a discussion held on May 21, 2022 at HKW in Berlin during the event *Unearthing the Present*.

Grün. auf ferne lassiret

C. No: 1 D

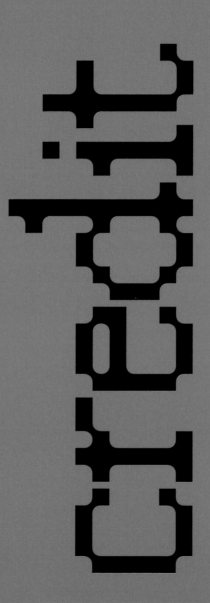

Slow Accumulation, Sudden Violence

Stalagmites and stalactites grow in caves, up from the floor and down from the ceiling. They are formed by water. Water in its normal room-temperature state is formless and flowing. But it is also a solvent. It carries things along with it, some of which have dissolved within it and some of which may perhaps ride on top of it or within it. And some of the things that are carried along with water can harden, ossify, and solidify over time.

The size, shape, and composition of speleothems (the general name for cave mineral deposits, of which stalactites and stalagmites are examples) are determined, in one sense, by very slow processes of accumulation that take place over long periods of time. These rates of accumulation can change as the environmental conditions change. The change of seasons brings changes in the chemical content of these structures, as does the change in the levels of carbon dioxide in the drip water that contributes to these stalagmites and stalactites. And these changes are reflected in very slow changes in the size, shape, and composition of the speleothems, changes that occur on the timescale of years, and larger changes on the timescale of centuries and millennia and even longer periods of time.

But the size, shape, and composition of these rock formations are also determined by extremely quick processes—processes including the kinds of interventions made by some of Earth's more commercially oriented creatures. I stumbled across a news article from October 2021 that covered one such very quick change in the height, size, and shape of some stalagmites in a cave in the Kosciuszko National Park in New South Wales, Australia.[1] Vandals had broken into the cave and cut down some of the stalagmites, most likely in order to take the minerals that had been congealed into these rock formations. So the kind of change which, at a growth rate of a millimeter per year, would have taken millennia upon millennia to make to these rock formations,

1   Sam McPhee, "Vandals break into an ancient cave in Kosciuszko National Park and cut up stalactites over two million years old" (October 25, 2021), https://www.dailymail.co.uk/news/article-10129959/Vandals-break-ancient-cave-Kosciuszko-National-Park-cut-stalactites-two-millions-years-old.html, accessed October 28, 2021.

happened in the space of a single evening because of the determinant
intervention of particular kinds of Earth creatures: humans, presum-
ably after money.

So why talk about this? Well, like the drip water that contributes
to the size, shape, and constitution of these rock formations, the normal course of our material and political circumstances as human beings is quite liquid. It's free flowing, rapidly changing; its fads and
fashions may settle in the pools of human historical contingency and
experience, staying pooled there for as long as a few years or perhaps
a few months, but soon flowing away.

However, some of the material carried in fads and fashions—
whether these fashions are literal fashions of aesthetics or whether
they are intellectual fashions or perhaps even moral fashions—con-
tain the kind of cultural material that can be recirculated, redistribut-
ed, and deposited by these chancier liquid flows of human contingency. This material may yet, like stalagmites and stalactites, solidify into
hard concrete institutions, traditions, and norms. History as we know
it is built on such material, cultural, and political solidifications and
accumulations.

All this has long been of interest to people from materialist
schools of philosophy. For example, in the famous speech "National
Liberation and Culture," the Bissau-Guinean revolutionary Amílcar
Cabral described history in much this way.[2] And he was responding to
a particular, very quick segment of history. The long periods of cultural, material, and political development that had been taking place
throughout many civilizations, including on African and American
continents, were very quickly—in what amounts to a single evening on
the scale of human history—altered by what I've come to call "global
racial empire," what others have called racial capitalism or European
colonialism. Whatever we call it, its violence caused very sudden and
irreparable changes in these accumulating and slowly built formations.

It's not the first time that this has happened. In fact, these kinds
of changes at the local level were endemic to much of human history.
What was different about the kinds of things that built the global racial
empire is the global part. Beginning in the fifteenth century, there was

---

2 Amílcar Cabral, "National Liberation and Culture," *Transition*, vol. 45
   (1974), pp. 12–17.

colonial conquest and soon thereafter slave trading, an accumulation of power and advantages on a planetary scale. The system built atop the global racial empire—that of planetary-scale trade, economic networks, and slave trading—has resulted in a cave formation that funnels liquid human advantages, social advantages, to those on top of our various hierarchies, forming the stalactites of wealth, state capacity, and research capacity for those in the Global North. And in the same way, it funnels accumulations of disadvantage towards those structurally on the bottom of global hierarchies, the stalagmites of poverty and of pollution.

It was these caves and the channels of hard political inertia and liquid circumstance and contingency that built the energy revolution and the industrial revolutions around it. It was the emissions around said revolutions that built the climate crisis. And it is these solid political structures that have accumulated throughout the new global racial empire—this built cave structure of politics—that stands between us and changing the things that we need to change in order to continue life on this planet on terms that even resemble justice.

What I've learned from the vandals at Kosciuszko National Park is a simple thing: the shape and stature of these long accumulations, stalagmites and stalactites, are simply the results of history. They are the results of a long history of accumulating drips of water, but they are also the results of a history of the inputs, the historical inputs of the activities of various creatures with the power to very quickly change these long-running accumulations.

A vandal with the right kind of tools can make the kinds of changes to caves that drip water makes, but very quickly. The difference and overlap of very different scales, between slower and quicker temporalities, has long been noticed by scholars. Slow and fast processes explain both the shape of stalagmites and stalactites and the vandalism at Kosciuszko National Park. And so it is with the climate crisis as well, which is affecting slow geological processes with the speed of cave vandals.

The question before us, before our own generation and the generations that are soon to come, is whether or not we will continue to allow fossil capital and its attendant capitalist structures to misuse this capacity for speed that human beings, as a very particular kind of creature with our social and economic networks, have. Whether we use that fast capacity to further destroy the conditions of life for us and for many of the Earth creatures besides us, or whether, with just as much speed and urgency, we will turn the same energies toward the project of changing our political and social conditions in order to harmonize rather than disrupt our relationships to the ecology around us.

The same capacities used by the vandals at Kosciuszko National Park have other uses, and it is within our power to direct those energies toward better ones. Will we? Only the various overlapping scales of time could possibly tell.

Nō: 29.

+   Katrin Klingan is a literary scholar, curator and producer of art and cultural projects. Between 2011 and 2022, she was curator at Haus der Kulturen and organized research projects, which explored the entanglement between human culture, natural environments, and global technologies, as well as structures of inequality and asymmetrical power relations. Together with Christoph Rosol she has headed the *Anthropocene Curriculum* since 2013, an international network and research project that explores, in an experimental and collaborative manner, pathways towards a new interdisciplinary culture of knowledge and education. Her recent projects at HKW include *Mississippi. An Anthropocene River* (2018–19), *Life Forms* (2019), *Unearthing the Present* and *Where is the Planetary?* (2022).

+   Giulia Rispoli is an assistant professor in History of Science at Ca' Foscari University of Venice, and a visiting scholar at the Max Planck Institute for the History of Science (MPIWG) in Berlin. She specializes in the history and epistemology of systems thinking especially in Russia, Europe, and North America. Her current research, "Planetary Genealogies. Historicizing the Anthropocene" contributes to the study and evaluation of the historical, epistemological, and scientific foundations of the Anthropocene as an Earth-system phenomenon. Other research interests include geoanthropology, the study of forgotten sources of environmental thinking, the Nuclear Winter, and the relationship between art, science, and literature. She recently co-curated with Christoph Rosol the online publication *Anthropogenic Markers: Stratigraphy and Context* (anthropocene-curriculum.org, 2022).

\+   **Christoph Rosol is research scholar at the Max Planck Institute for the History of Science (MPIWG) in Berlin, where he heads the research group Anthropocene Formations. Acting as a liaison between MPIWG and Haus der Kulturen der Welt where he has been an associate researcher and curator since 2012, he has co-developed and co-led a variety of research programs and interdisciplinary projects, amongst which is the *Anthropocene Curriculum*, a global platform for experimental research and education. Recent publications include the special double issue of *The Mississippi Papers* (*The Anthropocene Review*, 2021) and the online publication (with Giulia Rispoli) *Anthropogenic Markers: Stratigraphy and Context* (anthropocene-curriculum.org, 2022).**

\+   **Niklas Hoffmann-Walbeck was program assistant in the Literature and Humanities department at Haus der Kulturen der Welt (HKW) from 2017 to 2022, focusing on Anthropocene-related research projects, publications, and events. Most recently, he has served as scientific coordinator for HKW's *Evidence & Experiment* project. He works as a curator, author, and translator: together with Janek Müller he was artistic director of the "Heat Cold Devices" (2018–19) project, and his German translation of Robert Byron's *The Station* was published by Die Andere Bibliothek in 2020.**

+   Kat Austen creates artworks and music in her studios in Berlin and Seoul. She is a senior teaching fellow at University College London. In her artistic practice, she focuses on environmental issues—her work driven by a motivation to explore how to move towards a more socially and environmentally just future. Among other venues, Austen's art has been exhibited and performed at Wrocław Contemporary Museum, Fusion Festival, Headlands Center for the Arts, BOZAR Center for Fine Art, Changwon Biennale, and Ars Electronica Festival. Her work is held in private and public collections.

+   Nigel Clark is Professor of Human Geography at Lancaster University. In research that looks at the way Earth processes shape, perturb, and inspire social life, his current concerns include the pyrogeography of explosions and the evolution of human care. He is co-author, with Bronislaw Szerszynski, of *Planetary Social Thought: The Anthropocene Challenge to the Social Sciences* (2021).

+   Kristine L. DeLong is Associate Professor in the Department of Geography and Anthropology, Louisiana State University. She joined the Department of Geography and Anthropology in August 2009 after completing her PhD in Marine Science at the University of South Florida. Her research is focused on climate change during the past 130,000 years, primarily in the subtropical to tropical regions.

+   Anna Echterhölter is Professor of History of Science at the University of Vienna since 2018. She was an interim professor at Technical University and at Humboldt-University Berlin and held fellowships at the Max Planck Institute for the History of Science, Berlin (2008, 2015), and the German Historical Institute, Washington, DC (2016). She is a co-founder of *ilinx* magazine and acting editor of *Science in Context*. Research topics include monetary valuation and metrology, the social history of quantification, epistemic decolonization and the history of indigenous law in the Pacific.

+   Victor Galaz is Associate Professor in Political Science and Deputy Director of the Stockholm Resilience Centre at Stockholm University. His research explores the political challenges created by rapid global change in the Anthropocene and the societal challenges created by technological change. He is the author of *Global Environmental Governance, Technology and Politics: The Anthropocene Gap* (2014).

+   Susan Schuppli is an artist-researcher, Director of the Centre for Research Architecture at Goldsmiths, University of London, and an affiliate artist-researcher and Board Chair of the research agency Forensic Architecture. Her work examines material evidence from war and conflict, environmental disasters, and climate change. Published recently was her *Material Witness: Media, Forensics, Evidence* (2020).

+   Olúfẹ́mi O. Táíwò is Associate Professor of Philosophy at Georgetown University, Washington, DC, where he teaches Africana philosophy, social and political philosophy, and ethics. He is the author of two recent volumes (2022) *Reconsidering Reparations* and *Elite Capture: How the Powerful Took Over Identity Politics (And Everything Else)*.

+   Liz Thomas is a paleoclimatologist and head of the ice-core research group at the British Antarctic Survey. Specializing in abrupt climate change, Thomas produced the first comprehensive record of Antarctic surface mass balance (SMB) and quantified the contribution the twentieth-century has made to global sea levels. In addition, she leads several research groups and is a member of the Expert Group for the International Partnerships in Ice Core Science (IPICS).

+   Simon Turner is a senior research fellow in Geography
at University College London. He investigates the changing
composition of sediments, illustrating the range of human
activities that can be identified. His PhD was an investigation
of coastal wetlands in Sicily. He is the scientific coordinator for
the Anthropocene Working Group (AWG) and HKW collaborative
project to seek a global boundary stratotype section and point
(GSSP) for the Anthropocene.

+   Abraham Gottlob Werner (1749–1817) was a German geologist and mineralogist known for developing mineralogy as an
independent discipline. He is considered the founder of geog-
nosy, the study of the physical and mineralogical composition
of the Earth, and was the main proponent of Neptunism, the
now-discarded geological theory according to which all rocks
were derived from the waters of a vast primordial ocean. With
his doctrine of external characteristics, he developed one of the
world's first systematic classifications of minerals. From 1775
until his death, he taught and conducted research at the Freiberg
Mining Academy. The Werner Mountains on the Antarctic
Peninsula are named after him.

+   Jens Zinke is a professor in the department of Palaeobiol-
ogy at the University of Leicester. His focus is on understanding
the natural and anthropogenic impacts from climate change
and land use change on tropical coral reefs in the past, present,
and future. His work with other scientists crosses traditional
disciplinary boundaries to achieve novel research outcomes.

Carbon Aesthetics Group

+ Desiree Foerster is Assistant Professor in Media and Culture Studies at Utrecht University. She was a postdoctoral instructor at the Cinema and Media Department, University of Chicago, and graduated from the Institute for Arts and Media, University of Potsdam, with her thesis "Aesthetic Experience of Metabolic Processes." Foerster has conducted several research-creation projects together with artists, designers, and academics from Concordia University, Arizona State University, and the Institute of Experimental Design and Media Cultures (IXDM) in Basel. Her research interests are aesthetics, media ecologies, affective media, embodiment, phenomenology, process philosophy, and immersive environments.

+ Myriel Milićević is an artist, interaction designer, and professor in the Department of Design at the University of Applied Sciences Potsdam. She investigates spaces of impossibilities, realignments of perspectives, and coexistence with other beings. These explorations are mostly of a participatory and collaborative nature, taking form in practical-utopian models, processes, mappings, and stories: turning a city's energy leaks into power sources, drawing counter-cycles to the nitrogen spills in our landscapes, guiding butterflies to new meadows in the Rocky Mountains, playing at biopiracy through Crops and Robbers, or telling stories with people in the hills of Thailand.

+ Karolina Sobecka is an artist and researcher whose work centers on the relationship between environmental concerns and science and technology development. Her current projects explore the histories of ecology and their legacies in the contemporary formulations of carbon governance. Sobecka's artwork has been shown internationally and received numerous awards, including from Creative Capital, the New York Foundation for the Arts, and the Princess Grace Foundation. She is a PhD researcher at the Critical Media Lab (CML) in Basel and a Visiting Predoctoral Fellow at the Max Planck Institute for the History of Science.

+   **Alexandra Regan Toland** is an artist with a PhD in Landscape Planning. She is Associate Professor for Arts and Research at the Bauhaus University in Weimar, where she directs the practice-based PhD program for Art and Design. She has published widely on the topics of art and environment, especially in contexts of soil protection issues, air pollution, and urban ecology. In her artistic research practice, she explores the political issues of urban soils, vegetation, and air in the Anthropocene. Since 2022, Toland has been co-chair of the International Union of Soil Sciences (IUSS) Commission on the History, Philosophy, and Sociology of Soil Science.

+   **Clemens Winkler** is a design researcher currently employed at the Cluster of Excellence "Matters of Activity: Image Space Material," at Humboldt University of Berlin. His practice focuses on entrance points into symbolic operations with objects and spaces through ephemeral material processes on various scales, especially through air, vapor, and dust. He has taught and exhibited internationally and is currently a guest professor in the "Spiel und Objekt" Master's degree program at the University of Performing Arts Ernst Busch in Berlin. Here, with students, Winkler has been devising playful frameworks for negotiation around current societal concerns, energy and pollution governance, post-fossil futures, and digital media use on theater stages.

Colophon

Das Neue Alphabet (The New Alphabet) is a publication series by HKW (Haus der Kulturen der Welt).

The series is part of the HKW project *Das Neue Alphabet* (2019–2022), supported by the Federal Government Commissioner for Culture and the Media due to a ruling of the German Bundestag.

Series Editors: Detlef Diederichsen, Anselm Franke,
      Katrin Klingan, Daniel Neugebauer, Bernd Scherer
Project Management: Philipp Albers
Managing Editor: Martin Hager
Copy-Editing: Mandi Gomez, Hannah Sarid de Mowbray
Design Concept: Olaf Nicolai with Malin Gewinner,
      Hannes Drißner

Vol. 24: *Evidence Ensembles*
Editors: Christoph Rosol, Giulia Rispoli, Katrin Klingan,
      and Niklas Hoffmann-Walbeck
Coordination: Niklas Hoffmann-Walbeck
Contributors: Kat Austen, Carbon Aesthetics Group (Desiree
      Foerster, Myriel Milićević, Karolina Sobecka, Alexandra
      Regan Toland, Clemens Winkler), Nigel Clark, Kristine
      L. DeLong, Anna Echterhölter, Victor Galaz, Susan
      Schuppli, Olúfẹ́mi O. Táíwò, Liz Thomas, Simon Turner,
      Abraham Gottlob Werner, and Jens Zinke
Graphic Design: Malin Gewinner, Hannes Drißner,
      Markus Dreßen, Lyosha Kritsouk
DNA-Lettering (Cover): Gerrit Brocks
Type-Setting: Lyosha Kritsouk
Fonts: FK Raster (Florian Karsten), Suisse BP Int'l (Ian Party)
      Lyon Text (Kai Bernau)
Image Editing: ScanColor Reprostudio GmbH, Leipzig
Printing and Binding: Gutenberg Beuys Feindruckerei GmbH,
      Langenhagen

Colophon

The photos in this publication show a selection of "The External Characteristics Collection" – part of the Mineralogical Collection of Abraham Gottlob Werner (1749–1817). Photos: Susanne Paskoff, TU Bergakademie Freiberg.

The graph on p. 65 is based on Stephen J. Pyne, *The Pyrocene: How we Created an Age of Fire, and what Happens Next*. University of California Press: Oakland, CA, 2021 © all rights reserved UC Press. This version revised by Nathaniel LaCelle-Peterson, Max Planck Institute for the History of Science (MPIWG).

Published by:
Spector Books
Harkortstr. 10
01407 Leipzig
www.spectorbooks.com

© 2022 the editors, authors, artists, Spector Books

Distribution:
Germany, Austria: GVA Gemeinsame Verlagsauslieferung
        Göttingen GmbH & Co. KG, www.gva-verlage.de
Switzerland: AVA Verlagsauslieferung AG, www.ava.ch
France, Belgium: Interart Paris, www.interart.fr
UK: Central Books Ltd, www.centralbooks.com
USA, Canada, Central and South America, Africa:
        ARTBOOK | D.A.P. www.artbook.com
Japan: twelvebooks, www.twelve-books.com
South Korea: The Book Society, www.thebooksociety.org
Australia, New Zealand: Perimeter Distribution,
        www.perimeterdistribution.com